VOLUME	EDITOR-IN-CHIEF	PAGES	
20	CHARLES F. H. ALLEN	113	*Out of print*
21	THE LATE NATHAN L. DRAKE	120	*Out of print*
22	LEE IRVIN SMITH	114	*Out of print*
23	LEE IRVIN SMITH	124	*Out of print*
24	THE LATE NATHAN L. DRAKE	119	*Out of print*
25	THE LATE WERNER E. BACHMANN	120	*Out of print*
26	THE LATE HOMER ADKINS	124	*Out of print*
27	R. L. SHRINER	121	*Out of print*
28	H. R. SNYDER	121	*Out of print*
29	CLIFF S. HAMILTON	119	*Out of print*
Collective Vol. 3	A revised edition of Annual Volumes 20–29 E. C. HORNING, *Editor-in-Chief*	890	
30	THE LATE ARTHUR C. COPE	115	*Out of print*
31	R. S. SCHREIBER	122	
32	RICHARD T. ARNOLD	119	*Out of print*
33	CHARLES C. PRICE	115	
34	WILLIAM S. JOHNSON	121	
35	T. L. CAIRNS	122	
36	N. J. LEONARD	120	
37	JAMES CASON	109	
38	JOHN C. SHEEHAN	120	
39	MAX TISHLER	114	
Collective Vol. 4	A revised edition of Annual Volumes 30–39 NORMAN RABJOHN, *Editor-in-Chief*	1036	
40	MELVIN S. NEWMAN	114	
41	JOHN D. ROBERTS	118	
42	VIRGIL BOEKELHEIDE	118	
43	B. C. MCKUSICK	124	
44	WILLIAM E. PARHAM	131	
45	WILLIAM G. DAUBEN	118	
46	E. J. COREY	146	
47	WILLIAM D. EMMONS	140	
48	PETER YATES	164	

ORGANIC SYNTHESES

CONTRIBUTORS

J. F. Arens
Robert D. Bach
A. A. Baum
Paul Block, Jr.
S. Boatman
A. P. Boisselle
J. Bornstein
R. D. Burpitt
Donald J. Burton
J. I. G. Cadogan
Sohan S. Chaudhary
R. Donald Clark
Peter M. Collins
Francis J. Creegan
Philip A. Cruickshank
H. Dorn
D. E. Eitel
Edward U. Elam
W. Elser
A. M. Feldman
J. P. Ferris
Nikolaus Fischer
Lester Friedman
A. B. Galun
P. G. Gassman
E. Gelblum
R. Gompper
Marjorie M. Goodman
P. Glenn Gott
W. H. Gumprecht
Phillip J. Hamrick
T. M. Harris
Harold Hart
Robert H. Hasek
Charles R. Hauser
A. Henderson
Frank E. Herkes
J. G. Higgins
Jürgen Hocker
A. K. Hoffmann
Preben Jakobsen
M. J. Jorgenson
A. Kalir
F. E. Kaminski
Keizo Kitahonoki
Minerva F. Kormendy
Katsumi Kotera
Peter Kovacic
Richard M. Lange
Sven-olov Lawesson
Annemarie Liedhegener
Bernard Loev
Francis M. Logullo
R. K. Mackie
R. W. Mancuso
J. L. Marshall
Gerald L. Mayberry
Gerard J. Mikol
Emil J. Moriconi
William S. Murphy
Melvin S. Newman
G. Opitz
Knud Pedersen
Manfred Regitz
Glen A. Russell
R. A. Sanchez
John P. Schaefer
Edward E. Schweizer
Arnold H. Seitz
John C. Sheehan
P. K. Shenov
J. E. Shields
J. A. Skorcz
Heinz A. Staab
A. F. Thacher
J. G. Thweatt
W. Voskuil
Kurt Wendel
E. H. White
A. Zubek

ORGANIC SYNTHESES

″

AN ANNUAL PUBLICATION OF SATISFACTORY METHODS FOR THE PREPARATION OF ORGANIC CHEMICALS

VOLUME 48

1968

JOHN WILEY AND SONS, Inc.
NEW YORK · LONDON · SYDNEY · TORONTO

Library of Congress Catalog Card Number: 21–17747

SBN 471 97240 1

PRINTED IN THE UNITED STATES OF AMERICA

NOMENCLATURE

Preparations appear in the alphabetical order of common names of the compounds. For convenience in surveying the literature concerning any preparation through *Chemical Abstracts* subject indexes, the *Chemical Abstracts* indexing name for each compound is given as a subtitle if it differs from the common name used as the title.

SUBMISSION OF PREPARATIONS

Chemists are invited to submit for publication in *Organic Syntheses* procedures for the preparation of compounds that are of general interest, as well as procedures that illustrate synthetic methods of general utility. It is fundamental to the usefulness of *Organic Syntheses* that submitted procedures represent optimum conditions, and the procedures should have been checked carefully by the submitters, not only for yield and physical properties of the products, but also for any hazards that may be involved. Full details of all manipulations should be described, and the **range** of yields should be reported rather than the maximum yield obtainable by an operator who has had considerable experience with the preparation. For each solid product the melting-point **range** should be reported, and for each liquid product the **range** of boiling point and refractive index should be included. In most instances, it is desirable to include additional physical properties of the product, such as ultraviolet, infrared, mass, or nuclear magnetic resonance spectra, and criteria of purity such as gas chromatographic data. The methods of preparation or sources of the reactants should be described in notes, and the physical properties (such as boiling point, index of refraction, melting point) of the reactants should be included except where standard commercial grades are specified.

Procedures should be written in the style and format employed in the latest published volume of *Organic Syntheses*. Copies of the current style sheet may be obtained from the Secretary of the Editorial Board.

Beginning with Vol. 49, **Sec. 3., Methods of Preparation,** and **Sec. 4., Merits of the Preparation,** will be combined into a single new **Sec. 3., Discussion.** In this section should be described other practical methods for accomplishing the purpose of the procedure that have appeared in the literature. It is unnecessary to mention methods that have been published but are of no practical synthetic value. Those features of the procedure that recommend it for publication in *Organic Syntheses* should be cited (synthetic method of considerable scope, specific compound of interest not likely to be made available commercially, method that gives better yield or is less laborious than other methods, etc.). If possible, a brief discussion of the scope and limitations of the procedure as applied to other examples as well as a comparison of the method with the other methods cited should be included. If necessary to the understanding or use of the method for related syntheses, a brief discussion of the mechanism may be placed in this section. The present emphasis of *Organic Syntheses* is on model procedures rather than on specific compounds (although the latter are still welcomed), and the **Discussion** section should be written to help the reader decide whether and how to use the procedure in his own research. Three copies of each procedure should be submitted to the Secretary of the Editorial Board. It is sometimes helpful to the Board if there is an accompanying letter setting forth the features of the preparation that are of interest.

Additions, corrections, and improvements to the preparations previously published are welcomed and should be directed to the Secretary.

EDITOR'S PREFACE

The preparations in this volume, like those in earlier volumes of *Organic Syntheses,* fall into two major categories. In one category are procedures that illustrate general synthetic methods of importance and novelty; while in the other are procedures for the preparation of compounds of specific interest in many fields of organic chemistry. The trend in recent volumes toward an emphasis on the first category, maintained in the present volume, reflects the continuing development of new general synthetic methods in the current remarkable upsurge of synthetic organic chemistry.

In this first category are procedures that illustrate the conversion of alcohols to halides by triphenylphosphine-halogen adducts (cinnamyl bromide), preparation of phenols from alkyl halides without cinesubstitution (3-hydroxypyrene), allylic oxidation with *t*-butyl perbenzoate (3-benzoyloxycyclohexene), conversion of tertiary hydrocarbons by nitrogen trichloride to *t*-alkylamines (1-amino-1-methylcyclohexane), preparation of amines by reductive cleavage of sulfonamides by hydrobromic acid in the presence of phenol (1,3-dihydroisoindole), reductive dechlorination of polychloro compounds with sodium and *t*-butyl alcohol in tetrahydrofuran (7,7-dimethoxybicyclo[2.2.1]-heptene), preparation of α-keto aldehydes via treatment of esters with the potassium salt of dimethyl sulfoxide (phenylglyoxal), synthesis of diazo compounds from compounds with active methylene groups by reaction with *p*-toluenesulfonyl azide (*t*-butyl diazoacetate), preparation of carbamates from alcohols with sodium cyanate and trifluoroacetic acid (*t*-butyl carbamate), preparation of carbodiimides by dehydration of ureas with *p*-toluenesulfonyl chloride and triethylamine [1-ethyl-3-(3-dimethylamino)propylcarbodiimide], selective alkylation of ketones via the dianion of their formyl derivatives (2-*n*-butyl-2-methylcyclohexanone), preparation of 1,1-diphenyl-substituted hydrocarbons by alkylation of diphenylmethane (1,1-diphenylpentane), preparation of diarylmethanes via reaction of aromatic aldehydes with chloral (*p*-bromodiphenylmethane), synthesis of cyclic ketones by ring expansion of their lower homologs via enamine formation and cycloaddition of ethyl propiolate (cyclodecanone), conversion of cinnamic acid and its derivatives to phenylcyclopropanes with lithium aluminum hydride (1,1-diphenylcyclopropane), preparation of arenediazonium-2-carboxylates and their conversion to biphenylenes (benzenediazonium-2-carboxylate and biphenylene), preparation of 1-substituted benzocyclobutenes by addition of a side-chain carbanion to an aryne bond (1-cyanobenzocyclobutene), oxidation of polyalkylbenzenes to 2,4-cyclohexadienones by peroxytrifluoroacetic acid (2,3,4,5,6,6-hexamethyl-2,4-cyclohexadien-1-one), reductive cyclization of oximes to aziridines (*cis*-2-benzyl-3-phenylaziridine), synthesis of 2-hydroxyisocarbostyrils via nitrosation of 1-indanones (2-hydroxy-3-methylisocarbostyril), reductive cyclization of nitro compounds to nitrogen heterocycles by triethyl phosphite (2-phenylindazole), preparation of ketene S,N-acetals from thioamides (2-methylmercapto-N-methyl-Δ^2-pyrroline), and the preparation of 1-alkyl-3-aryltriazenes and their use in the esterification of acids (1-methyl-3-*p*-tolyltriazene).

Procedures for the preparation of several compounds of considerable utility are described. These include 1,1′-carbonyldiimidazole, which has been used in the preparation of esters, amides, and anhydrides, the hydrochloride and methiodide of 1-ethyl-3-(3-dimethylamino)-propylcarbodiimide, which can be used for similar purposes and are especially useful in the preparation of peptides, and (+)- and (−)-α-(2,4,5,7-tetranitro-9-fluorenylideneaminooxy) propionic acid (TAPA), which is used for the resolution of polycyclic aromatic compounds.

The members of the Editorial Board take this opportunity to thank the contributors of preparations. They welcome suggestions of changes that would improve the usefulness of *Organic Syntheses*. The attention of submitters of preparations is particularly drawn this year to the instructions on pages v and vi, which reflect changes that will be introduced in the next volume.

PETER YATES

CONTENTS

AMINOMALONONITRILE *p*-TOLUENESULFONATE

(Malononitrile, amino-, *p*-toluenesulfonate)

$$CH_2(CN)_2 + NaNO_2 + CH_3COOH \rightarrow HON{=}C(CN)_2 + CH_3CO_2Na + H_2O$$

$$3HON{=}C(CN)_2 + 4Al + 9H_2O \rightarrow 3NH_2CH(CN)_2 + 4Al(OH)_3$$

$$NH_2CH(CN)_2 + p\text{-}CH_3C_6H_4SO_3H \rightarrow p\text{-}CH_3C_6H_4SO_3^- \overset{+}{N}H_3CH(CN)_2$$

Submitted by J. P. Ferris, R. A. Sanchez, and R. W. Mancuso [1]
Checked by O. W. Webster and R. E. Benson

1. Procedure

A. *Oximinomalononitrile.* Malononitrile (Note 1) (25 g., 0.38 mole) is dissolved in a mixture of 20 ml. of water and 100 ml. of acetic acid in a 1-l. round-bottomed flask equipped with a stirrer, a thermometer, and a powder funnel. The solution is cooled to −10° with a dry ice-acetone bath, and 50 g. (0.72 mole) of granulated sodium nitrite is added in approximately 2-g. portions over a 30-minute period while the temperature is maintained at 0° to −10°. After the addition is complete a wet ice bath is used to maintain the temperature below 5° while the mixture is stirred for 4 hours. Four hundred milliliters of tetrahydrofuran (Note 2) and 400 ml. of ether are added in separate portions, and the mixture is stored at −40° overnight. The mixture is filtered rapidly, and the solid is washed with a mixture of 200 ml. of tetrahydrofuran (Note 2) and 200 ml. of ether. The filtrate and washings are combined and concentrated by distillation to a volume of 250 ml. by the use of a water aspirator and a bath at 40° (Note 3). This solution of oximinomalononitrile is used directly in the next step.

B. *Aminomalononitrile p-toluenesulfonate.* Aluminum foil (13.7 g., 0.51 g. atom) is cut into half-inch squares and is covered with a 5% aqueous solution of mercuric chloride until a mercury coating is visible on the aluminum (*ca.* 30 seconds). The mercuric chloride solution is decanted, and the amalgamated aluminum is washed twice with water, once with ethanol, and twice with tetrahydrofuran (Note 2). The amalgamated aluminum is transferred to a 2-l. round-bottomed flask fitted with a condenser, a stirrer, and a 250-ml. addition funnel and is covered immediately with 300 ml. of tetrahydrofuran (Note 2). The mixture is cooled in a dry ice-acetone bath, and the solution of oximinomalononitrile from procedure A is added with stirring over a 15-minute period while the temperature is maintained at −15° to −30°. Stirring is continued for an additional 5 minutes. The dry-ice acetone bath is then removed, and the mixture is allowed to warm to room temperature. (*Caution! Cooling with a dry ice-acetone bath is usually needed to control the reaction.*) After the spontaneous reaction subsides, the mixture is warmed to reflux until most of the aluminum is consumed (45 minutes). The reaction mixture is cooled to room temperature, 200 ml. of ether is added with stirring, and the aluminum salts are removed by vacuum filtration through Celite filter aid. The solid is washed with 250 ml. of tetrahydrofuran (Note 2) followed by 500 ml. of ether (Notes 3 and 4). The original filtrate and washings are combined and concentrated to about 250 ml. by the use of a water aspirator and a bath at 40°. To the resulting brown solution is slowly added with stirring a mixture of 60 g. (0.32 mole) of *p*-toluenesulfonic acid monohydrate as a slurry in 250 ml. of ether (Note 5). The total volume is brought to 1 l. with ether, the mixture is cooled to 0°, and the crystalline solid is collected by vacuum filtration. The product is washed successively with 200 ml. of ether, 200 ml. of cold (0°) acetonitrile, and 200 ml. of ether and dried at 25° (1 mm.) to give light tan crystals, m.p. 169–171° (dec.); yield, 75–79 g. (78–82%).

This product is suitable for most synthetic purposes. An almost colorless product may be obtained by recrystallization from boiling acetonitrile (100 ml. dissolves 1.8 g. of product) with treatment with activated carbon. The recovery of aminomalononitrile *p*-toluenesulfonate, m.p. 172° (dec.), is *ca.* 80%.

2. Notes

1. Commercial malononitrile is purified by dissolving 260 g. in 1 l. of ether, refluxing the solution with 5 g. of activated carbon for 10 minutes, and filtering through Celite under vacuum. The malononitrile crystallizes from the filtrate as a result of the cooling and concentration during the filtration. It is collected by filtration and washed with 350 ml. of cold (−20°) ether to give 214 g. of white crystals.

2. Tetrahydrofuran from Fisher Scientific Co. was used by the checkers. [*Caution! See Org. Syntheses,* **46**, *105 (1966), for a warning regarding the purification of tetrahydrofuran.*]

3. Occasionally a precipitate may form in the filtrate. It is removed by filtration before proceeding to the next step.

4. Additional washing is necessary if the washings are not colorless at this point.

5. One can check for complete precipitation of the aminomalononitrile by adding *p*-toluenesulfonic acid to the clear supernatant liquid.

3. Methods of Preparation

The present procedure is a modification of the original synthesis.[2] Previous reports of the synthesis of aminomalononitrile are in error.[2] Oximinomalononitrile was prepared by a modification of the procedure of Ponzio.[3]

4. Merits of the Preparation

This procedure provides a convenient synthesis of aminomalononitrile, which has been demonstrated to be a useful intermediate for the preparation of substituted imidazoles, thiazoles, oxazoles, purines, and purine-related heterocycles.[2] It is also a convenient starting material for the preparation of diaminomaleonitrile.[2, 4]

1. The Salk Institute for Biological Studies, San Diego, California [Present address (J.P.F.): Department of Chemistry, Rensselaer Polytechnic Institute, Troy, New York 12181].
2. J. P. Ferris and L. E. Orgel, *J. Am. Chem. Soc.*, **88**, 3829 (1966); **87**, 4976 (1965).
3. G. Ponzio, *Gazz. Chim. Ital.*, **61**, 561 (1931).
4. J. P. Ferris and R. A. Sanchez, *Org. Syntheses*, this volume, p. 60.

1-AMINO-1-METHYLCYCLOHEXANE

(Cyclohexylamine, 1-methyl-)

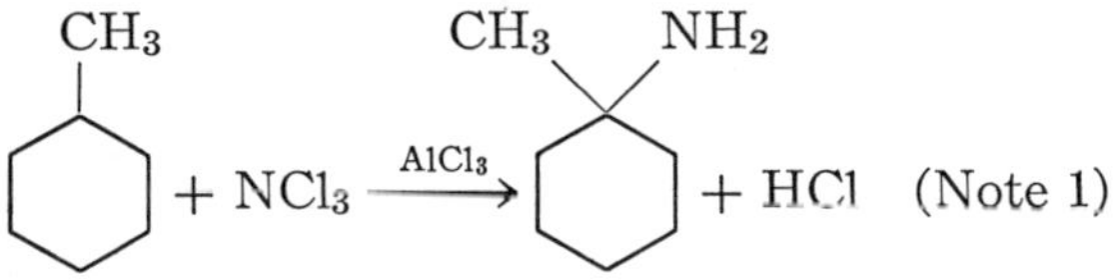

Submitted by Peter Kovacic and Sohan S. Chaudhary [1]
Checked by R. A. Haggard and William D. Emmons

1. Procedure

Caution! The reactions should be carried out in a hood behind a protective screen since trichloramine is noxious and potentially explosive; however, no difficulties from decomposition have been encountered under the conditions described.

A. *Trichloramine.* A mixture of 600 ml. of water (Note 2), 900 ml. of methylene chloride (Note 3), and 270 g. (1.32 moles) of calcium hypochlorite (Note 4) is cooled to 0–10° in a 3-l., three-necked, vented flask equipped with a stirrer, a thermometer, and a dropping funnel. A solution of 66.0 g. (1.23 moles) of ammonium chloride in 150 ml. of concentrated hydrochloric acid and 450 ml. of water is added dropwise with stirring over a 1-hour period at 0–10°. After an additional 20 minutes of stirring, the organic layer is separated, washed with three 200-ml. portions of cold water, and dried over anhydrous sodium sulfate. The yellow solution is filtered, and the trichloramine concentration is determined by iodometric titration (Note 5).

B. *1-Amino-1-methylcyclohexane.* A 3-l. three-necked flask is fitted with a paddle stirrer, a condenser, a thermometer, and a dropping funnel with an extension for below-surface addition. Provision is made for introduction of nitrogen by use of a side-arm adapter. The vessel is charged with 196 g. (2.0 moles) of methylcyclohexane (Note 6) and 106 g. (0.80 mole) of anhydrous aluminum chloride. A solution (*ca.* 600 ml.) of trichloramine (0.40 mole) in methylene chloride is added with efficient stirring over a period of 2 hours at −5° to 5° (Note 7). Throughout the reaction a stream of nitrogen is passed through the flask (Note 8). The brown mixture is stirred for an additional 20–30 minutes at the same temperature.

The reaction mixture is then added with good stirring to a slurry of 800–900 g. of ice and 50 ml. of concentrated hydrochloric acid (Note 9). The layers are separated, and the dark organic layer is washed with three 100-ml. portions of 5% hydrochloric acid and discarded. Traces of non-basic organic material are removed from the combined aqueous layer and washings by extraction with pure ether (Note 10) until the extract is colorless. The aqueous solution is treated with 600 ml. of 50% aqueous sodium hydroxide (Note 11) with cooling, and the basic organic product is extracted with three 125-ml. portions of pure ether (Note 10). The ethereal solution is dried over sodium sulfate, and the solvent is distilled on the steam bath to give 42–46 g. of a clear, amber product (Note 12). To this crude product is added 10 g. of triethylenetetramine (Note 13). Distillation through a small Vigreux column yields 21.5–30 g. (48–67%, based on trichloramine) of 1-amino-1-methylcyclohexane, b.p. 44–49° (20–25 mm.), $n^{22}\text{D}$ 1.4516 (Note 14).

2. Notes

1. The stoichiometry of the reaction is not known.
2. Deionized water is used throughout.
3. Commercial methylene chloride was distilled before use by the submitters. The checkers used reagent grade methylene chloride without distillation.
4. Calcium hypochlorite is obtained as "HTH" (Olin Mathieson Chemical Co., 70% purity).
5. Iodometric determination of positive chlorine is carried out

as follows: 2.0 g. of potassium iodide or sodium iodide is dissolved in 10 ml. of water, and 40 ml. of glacial acetic acid is added. Into this solution is pipetted 1.0 ml. of the methylene chloride solution of trichloramine. The liberated iodine is titrated with 0.100*N* sodium thiosulfate. The solution is found to be 0.6–0.7*M* in trichloramine. Storage for several days at 0–5° results in negligible decomposition, although it is not recommended unless adequate safety precautions are observed. Excess methylene chloride-trichloramine solution can be conveniently disposed of by its slow addition to a cold, stirred, dilute aqueous solution of sodium metabisulfite.

6. A pure grade of methylcyclohexane (Eastman Organic Chemicals) is used. Subsequent to the checking of this preparation, the submitters reported 69–72% yields with 78.4 g. (0.80 mole) of methylcyclohexane.[2] In this case a 1-l. three-necked flask is employed for the reaction; the remainder of the procedure is unchanged.

7. Cooling is accomplished with either an ice-salt bath or preferably a dry ice-acetone bath. The time of addition can be reduced to 1 hour by use of the latter. However, if the temperature is much below that designated, unchanged trichloramine accumulates, resulting eventually in an uncontrollable reaction.

8. Purging with nitrogen results in some increase in yield. If the flow is too vigorous, trichloramine is lost by volatilization.

9. The mixture can be stored overnight at this stage.

10. High-purity ether (*e.g.*, Baker Analyzed Reagent) is used since a grade of lower quality gives a product that is more difficult to purify because of contamination with alcohol.

11. Excess sodium hydroxide is needed to dissolve the aluminum-containing precipitate.

12. The last portion of solvent is carefully removed at the water aspirator.

13. Triethylenetetramine (redistilled, Eastman Organic Chemicals) prevents bumping and foaming and acts as a chaser for the distillation.

14. The product contains less than 10% of lower-boiling impurities determined (by the checkers) by vapor-phase chromatography with a column packed with 15% XF-1150 on Chromo-

sorb W. Further purification can be effected readily with good recovery by drying over sodium hydroxide pellets and fractionating at atmospheric pressure through an efficient spinning band column, with collection of the fraction, b.p. 142–146°, n^{22}D 1.4522.

3. Methods of Preparation

In addition to the present method,[2] 1-amino-1-methylcyclohexane has been synthesized by the following procedures: Ritter reaction, *e.g.*, with 1-methylcyclohexanol (76%, 67%)[3, 4] or 1-methylcyclohexene (35%);[4] Hofmann reaction with 1-methylcyclohexanecarboxamide (80% as hydrochloride);[5] reduction of 1-methyl-1-nitrocyclohexane (63%);[5] Schmidt reaction with 1-methylcyclohexanecarboxylic acid (42%).[6]

4. Merits of the Preparation

This procedure constitutes the first example of one-step conversion of a *t*-alkane to the corresponding *t*-alkylamine. Other hydrocarbons in this class, such as isobutane, have also been aminated with good results.[7] Only a very limited number of convenient routes, *e.g.*, the Ritter reaction, are available for the preparation of *t*-carbinamines. The present preparation illustrates a simple method that utilizes a novel substrate.

1. Department of Chemistry, Case Western Reserve University, Cleveland, Ohio 44106.

2. P. Kovacic and S. S. Chaudhary, *Tetrahedron*, **23**, 3563 (1967).
3. H. J. Barber and E. Lunt, *J. Chem. Soc.*, 1187 (1960).
4. W. Haaf, *Ber.*, **96**, 3359 (1963).
5. K. E. Hamlin and M. Freifelder, *J. Am. Chem. Soc.*, **75**, 369 (1953).
6. C. Schuerch, Jr., and E. H. Huntress, *J. Am. Chem. Soc.*, **71**, 2233 (1949).
7. P. Kovacic and S. S. Chaudhary, unpublished work.

3(5)-AMINOPYRAZOLE

[Pyrazole, 3(or 5)-amino-]

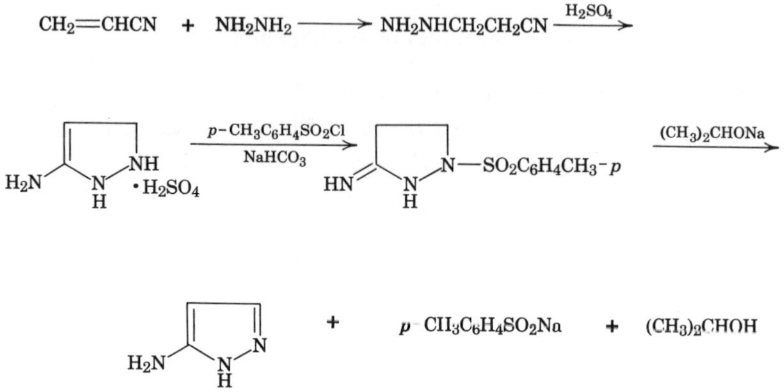

Submitted by H. DORN and A. ZUBEK [1]
Checked by L. G. VAUGHAN and R. E. BENSON

1. Procedure

A. *β-Cyanoethylhydrazine.* To a 2-l. two-necked flask fitted with a thermometer and a pressure-equalizing funnel are added a large magnetic stirring bar and 417 g. (6.00 moles of $N_2H_4 \cdot H_2O$) of 72% aqueous hydrazine hydrate. Acrylonitrile (318 g., 6.00 moles) is gradually added with stirring during 2 hours. The internal temperature is kept at 30–35° by occasional cooling of the flask. The funnel is replaced by a distillation condenser. Removal of water by distillation at 40 mm. at a bath temperature of 45–50° gives 490–511 g. (96–100%) of β-cyanoethylhydrazine as a yellow oil that is suitable for use in the next step. This product can be purified by distillation; b.p. 76–79° (0.5 mm.).

B. *3-Amino-3-pyrazoline sulfate.* In a 2-l. four-necked flask equipped with a reflux condenser, a dropping funnel, a thermometer, and a mechanical stirrer with four blades (Note 1) is placed 308 g. (169 ml., 3.0 moles) of 95% sulfuric acid (sp. gr. 1.834). Absolute ethanol (450 ml.) is added dropwise over 20–30 minutes. The internal temperature is maintained at 35° by cooling. A solution of 85.1 g. (1.00 mole) of β-cyanoethylhydrazine in 50 ml. of absolute ethanol is added with vigorous stirring over 1–2 minutes without further cooling (Note 1). The mixture warms spontaneously to 88–90° and is kept at this temperature for 3 minutes until the product begins to crystallize. The temperature of the stirred mixture is gradually lowered during the next hour to 25° by cooling with water, and the mixture is then allowed to stand at room temperature for 15–20 hours. The crystals are collected by filtration and washed three times with 80 ml. of absolute ethanol and finally with 80 ml. of ether. After being dried at 80° the product weighs 177–183 g. (97–100%), m.p. 143–144° (Note 2). The product is sufficiently pure for use in the following step; it may be recrystallized from methanol to give white needles, m.p. 144–145° (Note 2).

C. *3-Imino-1-(p-tolylsulfonyl)pyrazolidine.* To a 3-l. four-necked flask fitted with a condenser, a thermometer, a wide-mouthed funnel, and a high-speed mechanical stirrer having five pairs of blades are added 183 g. (1.00 mole) of 3-amino-3-pyrazoline sulfate and 1 l. of water. Sodium bicarbonate (210 g., 2.5 moles) is gradually added during 10 minutes with stirring. The rate of stirring is increased to 5000–6000 r.p.m., and a solution of 229 g. (1.20 moles) of *p*-toluenesulfonyl chloride in 400 ml. of benzene containing 0.5 g. of sodium dodecylbenzenesulfonate (Note 3) is added at one time. Three further portions of sodium bicarbonate are added sequentially: 25.2 g. (0.30 mole) after 15 minutes; 16.8 g. (0.20 mole) after 30 minutes; 16.8 g. (0.20 mole) after 55 minutes. The mixture is stirred for 5 hours at 18–25°, occasional cooling being required. Sodium bicarbonate (8.4 g., 0.10 mole) is added, then 200 ml. of ether, and stirring is continued for another hour. The colorless product is collected by filtration on a sintered-glass funnel, washed with three 50-ml. portions of ether followed by 50 ml. of water, and dried at 90°. The yield is 139–180 g. (58–75%); m.p. 183–185° (Note 4). The product is used directly in the next step.

D. *3(5)-Aminopyrazole* (Note 5). (*Caution! Because hydrogen gas is evolved, this reaction should be conducted in an efficient hood in the absence of an ignition source.*) A solution of sodium isopropoxide is prepared from 18.4 g. (0.80 g. atom) of sodium and 500 ml. of isopropyl alcohol in a 2-l. four-necked flask fitted with a mechanical stirrer, a thermometer, a reflux condenser, and a stopper. The reflux condenser is fitted with a nitrogen-inlet line attached to a bubbler device to maintain an anhydrous atmosphere. After all the sodium has dissolved, the temperature is adjusted to 60–70°, the stopper is replaced by a wide-mouthed funnel, and 191 g. (0.80 mole) of 3-imino-1-(*p*-tolylsulfonyl)-pyrazolidine is added gradually over 10 minutes to the hot solution under a blanket of nitrogen. The funnel is replaced by the stopper, and the mixture is stirred vigorously and then refluxed briefly. Stirring is continued, and the mixture is allowed to cool to room temperature during 2 hours. The precipitated sodium *p*-toluenesulfinate (140–142 g.) is removed by filtration and washed with a total of 100 ml. of isopropyl alcohol in several portions. The filtrate is treated twice with 4-g. portions of Norit activated carbon. The solvent is removed by distillation, the final trace being removed at a bath temperature of 50° (20 mm.) to give 62–66 g. (93–99%) of 3(5)-aminopyrazole as a light yellow oil. This is purified by distillation to give the product as a yellow oil, b.p. 100–102° (0.01 mm.), in 74–84% recovery (Note 6). The product crystallizes on cooling; m.p. 37–39° (Note 7). Its n.m.r. spectrum (60 MHz, dimethyl sulfoxide-d_6) shows two one-proton doublets at δ 7.33 and 5.52 p.p.m. (J = 2 Hz) and a broad three-proton singlet at δ 7.05 p.p.m. that is absent after addition of D_2O.

2. Notes

1. A stirrer with large blades operating at high speed is essential. Inadequate stirring results in solidification of the reaction mixture and makes proper washing of the product very difficult.

2. The checkers found melting points of 138–141° and 140–142°. After three recrystallizations from methanol the product has a melting point of 139.7–140°. The product appeared to be unstable to prolonged heating in methanol.

3. This salt serves as an emulsifying agent.

4. A sample, m.p. 184–185°, prepared by recrystallization of the product from nitromethane, gives satisfactory elemental analytical data. Its n.m.r. spectrum (60 MHz, dimethyl sulfoxide-d_6) reveals that the compound exists in the iminopyra-

zolidine form under these conditions; signals at δ 7.72 p.p.m. (doublet, $J = 8.4$ Hz), 7.40 p.p.m. (doublet, $J = 8.4$ Hz), 6.1 p.p.m. (broad singlet; absent after addition of D_2O), 3.4 p.p.m. (triplet, $J = 9.0$ Hz), and 2.4 p.p.m. (sharp singlet superimposed on triplet) with relative intensities of 2:2:2:2:5. The signals at 7.72 and 7.40 p.p.m. are assigned to the four aromatic protons, that at 6.1 p.p.m. to the two N—H protons, that at 3.4 p.p.m. to one pair of methylene protons, and that at 2.4 p.p.m. to the second pair of methylene protons plus the protons of the methyl group.

5. 3(5)-Aminopyrazole may also be obtained by hydrolysis of 3-imino-1-(*p*-tolylsulfonyl)pyrazolidine with aqueous alkali. In this case the pyrazolidine (239 g., 1.00 mole) is added to a solution of 40 g. (1.0 mole) of sodium hydroxide in 250 ml. of water at 75°, the resulting solution is stirred briefly, and the water is removed at reduced pressure. 3(5)-Aminopyrazole is separated from the sodium *p*-toluenesulfinate by several extractions with isopropyl alcohol.

6. In order to obtain maximum recovery the submitters conducted the distillation of 120 g. of crude product for 7–10 hours.

7. The checkers observed b.p. 119–121° (1.0 mm.) and m.p. 34–37°.

3. Methods of Preparation

3(5)-Aminopyrazole has been prepared by a Curtius degradation of pyrazole-3(5)-carboxylic acid hydrazide,[2, 3] by saponification and decarboxylation of ethyl 3-aminopyrazole-4-carboxylate[4] obtained from ethyl ethoxymethylenecyanoacetate and hydrazine, and by the present procedure.[5, 6]

4. Merits of the Preparation

This procedure represents the most convenient synthesis of 3(5)-aminopyrazole. It employs readily available starting materials and gives excellent yields in all steps.[5, 6] *p*-Toluenesulfonyl chloride can be replaced by other arenesulfonyl chlorides. 3-Imino-1-arylsulfonylpyrazolidines can be alkylated with dimethyl sulfate or with alkyl *p*-toluenesulfonates in dimethylformamide to give salts of 1-alkyl-2-arylsulfonyl-5-amino-4-

pyrazolines from which arenesulfinate can be eliminated as described in procedure D. In this fashion 1-alkyl-5-aminopyrazoles can be easily prepared.[6]

1. Institut für Organische Chemie der Deutschen Akademie der Wissenschaften, Berlin-Adlershof, Germany.
2. L. Knorr, *Ber.*, **37**, 3520 (1904).
3. H. Reimlinger, A. van Overstraeten, and H. G. Viehe, *Ber.*, **94**, 1036 (1961).
4. P. Schmidt and J. Druey, *Helv. Chim. Acta*, **39**, 986 (1956).
5. H. Dorn, G. Hilgetag, and A. Zubek, *Angew. Chem.*, **76**, 920 (1964); *Angew. Chem. Intern. Ed. Engl.*, **3**, 748 (1964).
6. H. Dorn, G. Hilgetag, and A. Zubek, *Ber.*, **98**, 3368 (1965).

BENZENEDIAZONIUM-2-CARBOXYLATE

(Benzenediazonium, *o*-carboxy-, hydroxide, inner salt)

AND BIPHENYLENE

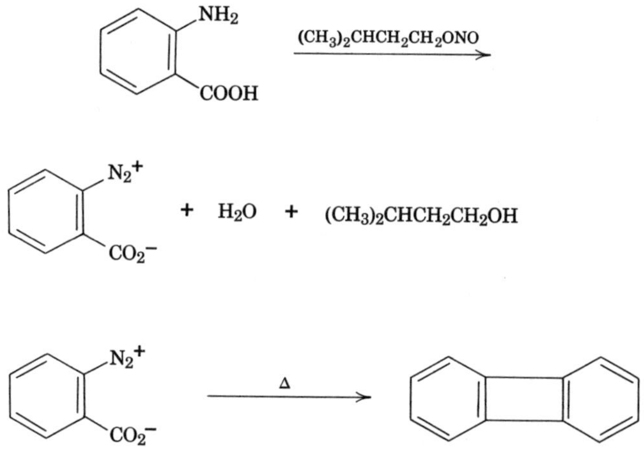

Submitted by FRANCIS M. LOGULLO, ARNOLD H. SEITZ, and LESTER FRIEDMAN[1]

Checked by G. D. ABRAMS, HERMANN ERTL, and PETER YATES

1. Procedure

Caution! Benzenediazonium-2-carboxylate when dry detonates violently on being scraped or heated, and it is strongly recommended that it be kept wet with solvent at all times. It should be prepared and used in a hood behind a safety screen. A wet towel or sponge should be kept within easy reach with which to deactivate any spilled material, which should then be disposed of by flooding with water.

A. *Benzenediazonium-2-carboxylate.* A solution of 34.2 g. (0.25 mole) of anthranilic acid (Note 1) and 0.3 g. of trichloroacetic acid (Note 2) in 250 ml. of tetrahydrofuran (Note 3) is prepared in a 600-ml. beaker equipped with a thermometer and cooled in an ice-water bath. The solution is stirred magnetically, and 55 ml. (48 g., 0.41 mole) of isoamyl nitrite (Note 4) is added over a period of 1–2 minutes. A mildly exothermic reaction occurs, and the reaction mixture is maintained at 18–25° and stirred for a further 1–1.5 hours. A transient orange to brick-red precipitate may appear (Note 5) which is slowly converted to the tan product. When the reaction is completed, the mixture is cooled to 10°, and the product is collected by suction filtration on a plastic Buchner funnel and washed on the funnel with cold tetrahydrofuran until the washings are colorless. (*Caution! The filter cake should not be allowed to become dry.*) The benzenediazonium-2-carboxylate is then washed with two 50-ml. portions of 1,2-dichloroethane to displace the tetrahydrofuran, and the solvent-wet material is used in the next step (Notes 6, 7, and 8).

B. *Biphenylene.* The solvent-wet benzenediazonium-2-carboxylate is washed from the funnel into a 400-ml. beaker with *ca.* 150 ml. of 1,2-dichloroethane, dispensed from a plastic wash bottle, with the aid of a plastic spatula (Note 9). The resultant slurry is added during 3–5 minutes to 1250 ml. of gently boiling, stirred 1,2-dichloroethane in a 2-l. beaker on a magnetic stirrer-hot plate in the hood (Note 10). Frothing ceases a few minutes after completion of the addition, and the mixture assumes a clear red-brown color, signaling the end of the reaction.

A 1-l., two-necked, round-bottomed flask is equipped with a 1-l. addition funnel and a Claisen distillation head and water-cooled condenser. The cooled reaction mixture is transferred to the funnel, and enough of it is admitted to the flask to half-fill the latter. The 1,2-dichloroethane, b.p. 83–84°, is distilled with the use of magnetic stirring to maintain even ebullition; the remainder of the reaction mixture is added from the funnel at a rate such that the flask remains about half-full. When *ca.* 75 ml. of dark residue remains in the flask, 300 ml. of ethylene glycol is added. An air condenser is substituted for the water-cooled condenser, and distillation is recommenced. A forerun, b.p. <150°, is discarded, and the fraction, b.p. 150–197°, is collected (Note 11). The distillate is cooled to 10°, and the product is collected by suction filtration, washed with 10–15 ml. of cold ethylene glycol and several times with water, and dried at atmospheric pressure over phosphorus pentoxide. The yield of biphenylene, m.p. 109–112°, is 4.0–5.6 g. (21–30%, based on anthranilic acid). Additional biphenylene (0.2–0.5 g.) can be obtained from the mother liquor and ethylene glycol washings by redistillation or dilution with water (Note 12).

2. Notes

1. The submitters used practical grade anthranilic acid from Mallinckrodt Chemical Works.

2. Perfluorobutyric or trifluoroacetic acid may be used in place of trichloroacetic acid. Strong mineral acids and acetic acid are wholly unsatisfactory. If a catalyst is not used, the product is of poor quality and the yield only 30%. Trichloroacetic acid is conveniently added as a solution in tetrahydrofuran (0.01 g./ml.).

3. The submitters used commercial tetrahydrofuran. The checkers found that the product yield was the same when either practical grade or Fisher Certified reagent grade tetrahydrofuran was used.

4. The submitters used "amyl nitrite" U.S.P. from Mallinckrodt Chemical Works; isoamyl nitrite supplied by Matheson, Coleman and Bell is apparently the same material. They found that other alkyl nitrites (ethyl, *n*-butyl, *t*-butyl, *n*-amyl) may be used with equal success. Subsequent to the checking of this procedure, they reported that the amount of nitrite can be reduced to a 20% molar excess.

5. This precipitate is apparently 2,2′-dicarboxydiazoaminobenzene.

6. The product should be used immediately because it deteriorates slowly at room temperature. It is freed of tetrahydrofuran by washing with the solvent to be used in subsequent reactions and transferred as a slurry in that solvent (cf. procedure B). Traces of water, if present, do not appear to interfere with subsequent reactions of diazonium carboxylates, as observed in the submitters' laboratory.

7. The submitters have found that this procedure works equally well with many substituted anthranilic acids; however, it does not work with 3-chloro-, 5-chloro-, 4-nitro-, 5-nitro-, and 4,5-benzoanthranilic acids.

8. Although it strongly recommended that the product not be dried, particularly when prepared on the scale described here, the following slightly modified procedure can be used for the preparation of solvent-free benzenediazonium-2-carboxylate. A solution of 2.74 g. (0.020 mole) of anthranilic acid and 0.030 g. of trichloroacetic acid in 30 ml. of tetrahydrofuran is prepared in a 100-ml. beaker equipped with a thermometer and cooled in a bath of *ca.* 25 g. of crushed ice. The solution is stirred magnetically, and 5 ml. (4.4 g., 0.038 mole) of isoamyl nitrite (Note 4) is added during *ca.* 0.5 minute. The mixture is stirred and allowed to warm to room temperature over a period of 1 hour. It is cooled to 10°, and the product is collected by suction filtration with the use of a plastic Buchner funnel and plastic spatula and washed with ice-cold tetrahydrofuran until the washings are colorless. The yield of air-dried (30 minutes) benzenediazonium-2-carboxylate is 2.55–2.88 g. (86–97%). (*Caution! Danger of detonation! See above.*) (Note 13).

9. The checkers transferred the benzenediazonium-2-carboxylate to the beaker with the aid of gentle air pressure (cf. Note 13) and then slurried it with *ca.* 150 ml. of 1,2-dichloroethane.

10. The checkers added the slurry via a large, medium-bore, glass funnel with fire-polished edges. In one of three runs a small, sharp report was heard, apparently from a source above the liquid in the beaker; the yield of biphenylene in this run did not differ significantly from that obtained in the other runs.

11. The checkers found it necessary to heat the condenser with a microburner from time to time to prevent clogging with

biphenylene. The submitters have reported that this can be avoided by connecting the Claisen head via an adapter to a two-necked receiving flask fitted with an upright water-cooled condenser and cooled by immersion in ice-water.

12. For convenient preparation and workup of larger amounts of biphenylene, several runs can be combined after the decomposition of the benzenediazonium-2-carboxylate. Thus the submitters obtained 19.1 g. (25%) of air-dried biphenylene by combining four batches. They found that the use of larger amounts of 1,2-dichloroethane resulted in a moderate increase in yield; by combining four batches, each prepared in 2.75 l. of 1,2-dichloroethane in a 4-l. beaker, they obtained 22.8 g. (30%) of product.

13. The checkers transferred the solvent-moist product to a tared Petri dish by means of a gentle puff of compressed air through the stem of the funnel; solid adhering to the filter paper and funnel was transferred to the dish with the aid of a soft rubber policeman, which was also used to spread the product over the surface of the dish. The product was then air-dried for 30 minutes in the hood.

3. Methods of Preparation

Benzenediazonium-2-carboxylate [2] and its substituted derivatives [3] have been prepared by diazotization of anthranilic acids in the presence of hydrochloric acid followed by dehydrochlorination of the resultant diazonium carboxylate hydrochlorides with silver oxide.

Biphenylene has been prepared in low yield by the reaction of 2,2′-dibromobiphenyl or 2,2′-biphenyliodonium iodide with cuprous oxide,[4] by the action of cupric chloride on 2,2′-biphenyldimagnesium dibromide,[5] from 2,2′-diiodobiphenyl via dibenzomercurole (*o,o*′-biphenylenemercury) (49%),[6] by pyrolysis or photolysis of phthaloyl peroxide (27%),[7] by reaction of *o*-fluorobromobenzene with lithium amalgam (24%),[8] by reaction of *o*-bromoiodobenzene with magnesium (12%),[9] and by the decomposition of diphenyliodonium-2-carboxylate,[10] 1,2,3-benzothiadiazole-1,1-dioxide,[11] and benzenediazonium-2-carboxylate.[12, 13]

4. Merits of the Preparation

These procedures illustrate facile methods for the preparation of benzenediazonium-2-carboxylate and its derivatives [14] and of biphenylene and certain biphenylene derivatives.[13] The latter preparation is far more convenient and proceeds in much better yield than do previous syntheses, which involve more steps, less accessible intermediates, and more complicated techniques.

1. Department of Chemistry, Case Western Reserve University, Cleveland, Ohio 44106.
2. A. Hantzsch and W. B. Davidson, *Ber.*, **29**, 1522 (1896).
3. M. Stiles, R. G. Miller, and U. Burckhardt, *J. Am. Chem. Soc.*, **85**, 1792 (1963).
4. W. C. Lothrup, *J. Am. Chem. Soc.*, **63**, 1187 (1941); W. Baker, M. P. V. Boarland, and J. F. M. McOmie, *J. Chem. Soc.*, 1476 (1954).
5. W. S. Rapson, R. G. Shuttleworth, and J. N. van Niekerk, *J. Chem. Soc.*, 326 (1943).
6. G. Wittig and W. Herwig, *Ber.*, **87**, 1511 (1954).
7. G. Wittig and H. F. Ebel, *Ann.*, **650**, 20 (1961).
8. G. Wittig and L. Pohmer, *Ber.*, **89**, 1334 (1956).
9. H. Heany, F. G. Mann, and I. T. Millar, *J. Chem. Soc.*, 3930 (1957).
10. E. Le Goff, *J. Am. Chem. Soc.*, **84**, 3786 (1962).
11. G. Wittig and R. W. Hoffmann, *Ber.*, **95**, 2718 (1962).
12. R. S. Berry, G. N. Spokes, and M. Stiles, *J. Am. Chem. Soc.*, **84**, 3570 (1962).
13. L. Friedman and A. H. Seitz, to be published.
14. L. Friedman, D. F. Lindow, and F. M. Logullo, to be published.

3-BENZOYLOXYCYCLOHEXENE

(2-Cyclohexen-1-ol, benzoate)

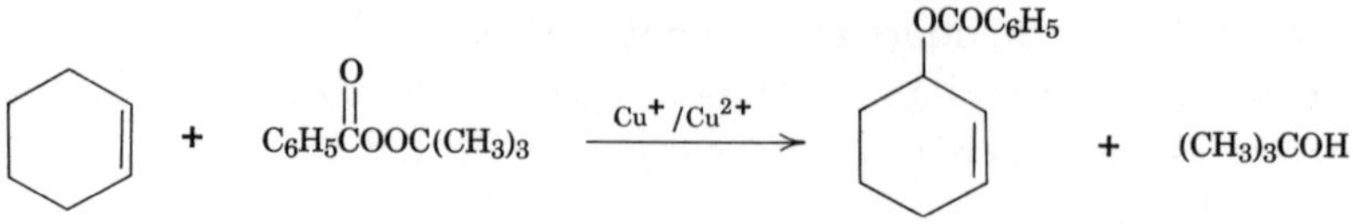

Submitted by KNUD PEDERSEN, PREBEN JAKOBSEN, and SVEN-OLOV LAWESSON [1]
Checked by R. SCHÖLLHORN and R. BRESLOW

1. Procedure

Caution! This reaction should be carried out behind a safety screen. The solvent removal and product distillation steps should also be carried out behind a screen to minimize danger due to contamination of the product with undetected peroxides.

A 250-ml., three-necked, round-bottomed flask equipped with a sealed mechanical stirrer, a reflux condenser, and a pressure-equalizing dropping funnel is set up for conducting a reaction in an atmosphere of nitrogen by fitting into the top of the condenser a T-tube attached to a low-pressure supply of nitrogen and to a mercury bubbler. In the flask are placed 41 g. (0.50 mole) of cyclohexene and 0.05 g. (0.00035 mole) of cuprous bromide, and the mixture is heated in an oil bath at 80–82°. When the temperature of the mixture reaches that of the oil bath, 40 g. (0.21 mole) of *t*-butyl perbenzoate (Note 1) is added dropwise with stirring over a 1-hour period, during which the color of the now homogeneous solution becomes blue. Stirring and heating are continued for an additional 3.5 hours (Note 2). The cooled reaction mixture is washed with two 50-ml. portions of dilute aqueous sodium carbonate to remove benzoic acid (Note 3). The remaining organic phase is washed with water until neutral and dried over anhydrous sodium sulfate. The excess of cyclohexene is removed by distillation under reduced pressure, and the residue (Note 4) is distilled through a short Vigreux column to give 29–33 g. (71–80%) of 3-benzoyloxycyclohexene, b.p. 97–99° (0.15 mm.), n^{20}_{D} 1.5376–1.5387 (Note 5).

2. Notes

1. *t*-Butyl perbenzoate is supplied by Lucidol Division, Wallace and Tiernan, Inc., Buffalo 5, New York, or L. Light and Co., Ltd., Colorbrook, Bucks, England. The Lucidol product contains 98% *t*-butyl perbenzoate.

2. The progress of the reaction can most conveniently be followed by periodic examination of the infrared spectrum of the mixture ($\nu_{C=O}$ for peroxybenzoate: 1775 cm.$^{-1}$). After all of the perester has been added, *ca.* 3 hours is required for its consumption.

3. After acidification of the aqueous phase 1.5–2 g. of benzoic acid can be isolated.

4. It is recommended that an infrared spectrum be run on the residue before the distillation to check for the absence of perester (see Note 2).

5. The same yield is obtained when the scale is increased threefold.

3. Methods of Preparation

The procedure is that of Kharasch, Sosnovsky, and Yang.[2]

4. Merits of the Preparation

The reaction described is of considerable general utility for the preparation of benzoyloxy derivatives of unsaturated hydrocarbons.[2-8] Reactions of *t*-butyl perbenzoate with various other classes of compounds in the presence of catalytic amounts of copper ions produce benzoyloxy derivatives. Thus this reaction can also be used to effect one-step oxidation of saturated hydrocarbons,[9, 10] esters,[5, 11] dialkyl and aryl alkyl ethers,[12-14] benzylic ethers,[11, 15] cyclic ethers,[13, 16] straight-chain and benzylic sulfides,[12, 17-19] cyclic sulfides,[11, 19] amides,[11] and certain organosilicon compounds.[20]

1. Department of Organic Chemistry, Chemical Institute, University of Aarhus, Aarhus C., Denmark.
2. M. S. Kharasch, G. Sosnovsky, and N. C. Yang, *J. Am. Chem. Soc.*, **81**, 5819 (1959).

3. D. B. Denney, D. Z. Denney, and G. Feig, *Tetrahedron Letters*, No. 15, 19 (1959).
4. A. L. J. Beckwith and G. W. Evans, *Proc. Chem. Soc.*, 63 (1962).
5. G. Sosnovsky and N. C. Yang, *J. Org. Chem.*, **25**, 899 (1960).
6. B. Cross and G. H. Whitham, *J. Chem. Soc.*, 1650 (1961).
7. J. K. Kochi, *J. Am. Chem. Soc.*, **84**, 774 (1962).
8. D. Z. Denney, A. Appelbaum, and D. B. Denney, *J. Am. Chem. Soc.*, **84**, 4969 (1962).
9. M. S. Kharasch and A. Fono, *J. Org. Chem.*, **23**, 324 (1958).
10. T. I. Wang, Ph.D. Thesis, University of Chicago, 1962.
11. C. Berglund and S.-O. Lawesson, *Arkiv Kemi*, **20**, 225 (1963).
12. S.-O. Lawesson, C. Berglund, and S. Grönwall, *Acta Chem. Scand.*, **15**, 249 (1961).
13. G. Sosnovsky, *Tetrahedron*, **13**, 241 (1961).
14. S.-O. Lawesson and C. Berglund, *Arkiv Kemi*, **17**, 465 (1961).
15. S.-O. Lawesson and C. Berglund, *Arkiv Kemi*, **16**, 287 (1960).
16. S.-O. Lawesson and C. Berglund, *Arkiv Kemi*, **17**, 475 (1961).
17. S.-O. Lawesson and C. Berglund, *Acta Chem. Scand.*, **15**, 36 (1961).
18. G. Sosnovsky, *Tetrahedron*, **18**, 15 (1962).
19. G. Sosnovsky, *Tetrahedron*, **18**, 903 (1962).
20. G. Sosnovsky and H. J. O'Neill, *Compt. Rend.*, **254**, 704 (1962).

cis-2-BENZYL-3-PHENYLAZIRIDINE

(Aziridine, 2-benzyl-3-phenyl-, *cis*-)

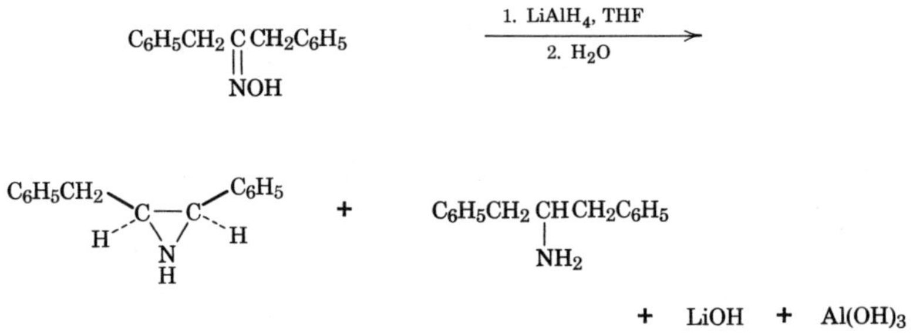

Submitted by Katsumi Kotera and Keizo Kitahonoki[1]
Checked by Donald R. Strobach and R. E. Benson

1. Procedure

In a 1-l., four-necked, round-bottomed flask fitted with a sealed mechanical stirrer, a thermometer, a dropping funnel, and a reflux condenser protected from atmospheric moisture with a drying tube containing calcium chloride are placed 350 ml. of dry tetrahydrofuran (Note 1) and 3.80 g. (0.100 mole) of powdered lithium aluminum hydride (Note 2). The slurry is stirred while a solution of 11.27 g. (0.0500 mole) of dibenzyl ketoxime (Note 3) in 80 ml. of dry tetrahydrofuran is added dropwise with cooling at 20° over a 10-minute period. The contents of the flask are gradually heated to reflux (Note 4) with stirring in an oil bath at 90° (external temperature) for 3 hours (Note 5); at *ca.* 62° the color of the mixture turns from the initial pale green to a permanent, light chocolate color (reaction may be exothermic at this point). The mixture is cooled with ice water and decomposed by gradual addition of 12 ml. of water at a temperature below 20°. The precipitate is collected by filtration, washed with 100 ml. of ether, and added to 200 ml. of ether. This mixture is stirred for *ca.* 10 minutes and filtered, and the residue is washed with 100 ml. of ether. The ethereal extracts and washings are combined with the original filtrate, dried over anhydrous sodium sulfate overnight, and concentrated with a rotary evaporator at 30° (20 mm.) to give 10.60–11.0 g. of a pale yellow oil (Note 6).

The product is dissolved in 100 ml. of petroleum ether, b.p. 30–40°, with warming, and the solution is transferred to a chromatographic column consisting of 75 g. of silica gel (Note 7). The product is eluted sequentially with (A) 300 ml. of petroleum ether, (B) 300 ml. of 3:1 (v/v) petroleum ether:benzene, (C) 300 ml. of 1:1 (v/v) petroleum ether:benzene, (D) 600 ml. of 1:3 (v/v) petroleum ether:benzene, and (E) 600 ml. of benzene. Fractions A and B are discarded (Note 8). The oil (8.50–9.15 g.) obtained by distillation of the solvent from the combined fractions C, D, and E is dissolved in 65 ml. of petroleum ether. Cooling gives 5.00–6.61 g. of colorless needles, m.p. 44–45° (Note 9). Concentration of the filtrate and cooling yield successive crops of product, m.p. 41–45°. The total yield is 7.45–8.15 g. (71–78%) (Note 9).

2. Notes

1. Tetrahydrofuran of laboratory chemical grade supplied by Fisher Scientific Co. was used without further purification by

the checkers. The submitters used tetrahydrofuran purified by the method of *Org. Syntheses*, Coll. Vol. **4**, 259 (1963). [*Caution! See Org. Syntheses*, **46**, *105* (*1966*), *for a warning regarding purification of tetrahydrofuran.*]

2. Obtained from Metal Hydrides, Inc.

3. The submitters used oxime prepared from Tokyo Kasei G. R. grade dibenzyl ketone in the usual manner and recrystallized from ether-petroleum ether; m.p. 123–124° (yield 93%).[2] The checkers prepared the oxime in the following manner. A mixture of 50 g. (0.24 mole) of 1,3-diphenyl-2-propanone (Eastman Organic Chemicals, practical grade), 50 g. (0.72 mole) of hydroxylamine hydrochloride, 250 ml. of reagent grade pyridine, and 250 ml. of ethanol was heated under reflux for 2 hours. The solvent was removed by distillation at reduced pressure, and the residue was triturated with 250 ml. of cold water. The solid was collected by filtration and washed with a small volume of cold water. Crystallization of the moist product from ethanol gave 50.5 g. (94%) of dibenzyl ketoxime, m.p. 122–124°.

4. The internal temperature is 66°. At lower temperatures the reaction takes longer, and the yield of the aziridine is lower. The submitters found that the yield is 66% after 6 hours at a reaction temperature of 50° and 55% after 30 hours at a temperature of 20° and 44 hours at −20°.

5. The consumption of the oxime can be checked by thin-layer chromatography on silica gel G with the solvent system chloroform/methanol (95/5 v/v) and a spray reagent consisting of 5% potassium dichromate in 40% sulfuric acid. The oxime appears as an immediate dark spot and the aziridine as a yellow spot. The checkers observed identical mobilities (R_f 0.8) for both compounds.

6. The submitters found that purification of the oil by direct crystallization gives only a small amount of the pure product. Attempted purification by distillation did not give satisfactory results.

7. Silica gel, particle size 0.2–0.5 mm. (Catalog No. 7733), of E. Merck A. G. (Darmstadt) was used.

8. The fractions are tested by thin-layer chromatography on

silica gel G with the solvent system and spray reagent described in Note 5.

9. The product is sufficiently pure for most purposes. The pure sample after additional recrystallizations melts at 44.7–45.1°.

3. Methods of Preparation

The only method reported[3] for the preparation of 2-benzyl-3-phenylaziridine is that described here.

4. Merits of the Preparation

The present preparation illustrates the general method for the synthesis of aziridines by reduction of ketoximes[3, 4] having an aromatic ring attached to carbon α or β to the oximino function and of aldoximes[4] having the aromatic ring attached to the carbon atom β to the oximino group. It has also been applied

TABLE I
AZIRIDINES PREPARED BY REDUCTION OF OXIMES WITH LITHIUM ALUMINUM HYDRIDE

Parent Ketone or Aldehyde	Aziridine	M.P., °C	Yield, %
Acetophenone	2-Phenylaziridine	(Oil)[a]	17
Phenylacetaldehyde	2-Phenylaziridine	(Oil)[a]	34
1-Acetonaphthone	2-(α-Naphthyl)-aziridine	66–67	64
3-Phenyl-2-butanone	2-(α-Methylbenzyl)-aziridine	(Oil)	38
1-Tetralone	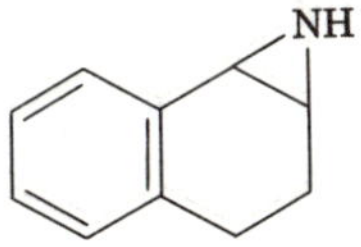	52–53.5[b]	11

[a] Cf. F. Wolfheim, *Ber.*, **47**, 1440 (1914); S. Gabriel and J. Colman, *Ber.*, **47**, 1866 (1914); S. J. Brois, *J. Org. Chem.*, **27**, 3532 (1962); A. Hassner and C. C. Heathcock, *Tetrahedron Letters*, 1125 (1964).

[b] Cf. G. Drefahl and K. Ponsold, *Ber.*, **93**, 519 (1960); A. Hassner and C. Heathcock, *Tetrahedron*, **20**, 1037 (1964).

to oximes of cyclic and bridged ring ketones, such as α- and β-tetralone,[3, 4] 1,2,3,4-dibenzo-1,3-cycloheptadien-6-one,[5] and bicyclo[2.2.2[octanone and its benzo analogs.[6] Examples of aziridines prepared by this method are given in Table I; derivatives of the products are listed in Table II. Because of the accessibility of oximes the present method provides a more convenient synthesis of several types of aziridines than do other methods.[7] Furthermore, the reaction proceeds stereoselectively to give the *cis*-substituted aziridine.[3]

TABLE II
DERIVATIVES OF AZIRIDINES PREPARED BY REDUCTION OF OXIMES WITH LITHIUM ALUMINUM HYDRIDE

Aziridine	1-(Phenylcarbamoyl) Derivative, M.P., °C	1-(*p*-Nitrobenzoyl) Derivative, M.P., °C	Derived Thiazolidine-2-thione, M.P., °C
2-Phenylaziridine		120–122.5	170–171[a,b] 168–169[b,c]
2-(α-Naphthyl)-aziridine	133.5–135		235–237 (dec.)
2-(α-Methylbenzyl)-aziridine		65–66 and 178–179[d]	96.5–97.5 and 165.5–166[d]
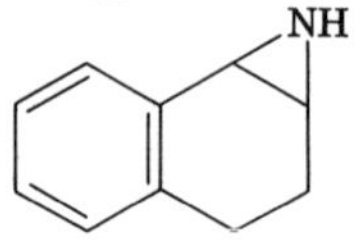	157–158		188.5–190.5

[a] Aziridine prepared from acetophenone.
[b] Cf. C. S. Dewey and R. A. Bafford, *J. Org. Chem.*, **30**, 491 (1965).
[c] Aziridine prepared from phenylacetaldehyde.
[d] Presumably *erythro* and *threo* isomers.

1. Shionogi Research Laboratory, Shionogi and Co., Ltd., Fukushima-ku, Osaka, Japan.
2. Cf. J. B. Senderens, *Bull. Soc. Chim. France*, [4] **7**, 645 (1910); C. H. DePuy and B. W. Ponder, *J. Am. Chem. Soc.*, **81**, 4629 (1959).
3. K. Kitahonoki, K. Kotera, Y. Matsukawa, S. Miyazaki, T. Okada, H. Takahashi, and Y. Takano, *Tetrahedron Letters*, 1059 (1965); M. Y. Shandala, M. D. Solomon, and E. S. Waight, *J. Chem. Soc.*, 892 (1965).
4. K. Kotera, S. Miyazaki, H. Takahashi, T. Okada, and K. Kitahonoki, *Tetrahedron*, **24**, 3681 (1968).

5. K. Kotera, M. Motomura, S. Miyazaki, T. Okada, and Y. Matsukawa, *Tetrahedron*, **24**, 1727 (1968).
6. K. Kitahonoki, Y. Takano, and H. Takahashi, *Tetrahedron*, **24**, 4605 (1968).
7. Cf. P. E. Fanta, in A. Weissberger, "Heterocyclic Compounds with Three- and Four-Membered Rings," Part 1, Interscience Publishers, New York, 1964, pp. 528–541.

BICYCLO[2.2.1]HEPTEN-7-ONE

(2-Norbornen-7-one)

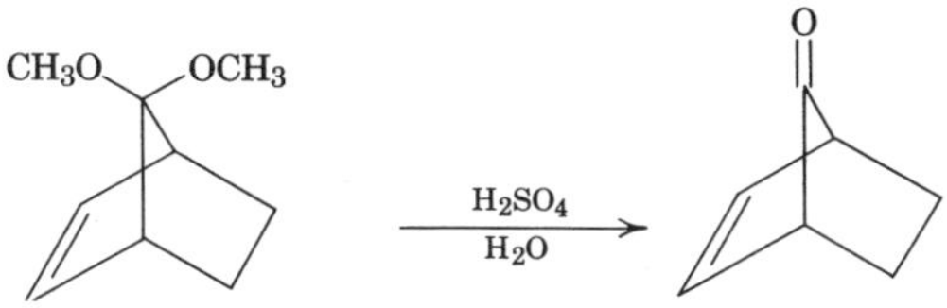

Submitted by P. G. Gassman and J. L. Marshall[1]
Checked by William G. Dauben and James L. Chitwood

1. Procedure

Into a 250-ml. Erlenmeyer flask are placed 45.9 g. (0.298 mole) of 7,7-dimethoxybicyclo[2.2.1]heptene (Note 1), 75 ml. of 5% aqueous sulfuric acid, and a Teflon-coated magnetic stirring bar. The flask is stoppered, and the mixture is stirred vigorously with a magnetic stirrer for 20 hours. The mixture is extracted with three 40-ml. portions of pentane, and the combined extracts are dried over anhydrous magnesium sulfate. The drying agent is removed by filtration, and the solvent is distilled through a 12-in. Vigreux column. Fractional distillation of the residual oil yields 28.9 g. (90%) of colorless bicyclo[2.2.1]hepten-7-one, b.p. 96–100° (115 mm.), n^{25}_{D} 1.4786 (Notes 2 and 3).

2. Notes

1. The preparation of 7,7-dimethoxybicyclo[2.2.1]heptene is described on p. 68.

2. The checkers, working at half-scale, obtained an 85% yield of product, b.p. 93–97° (118 mm.).

3. This material is extremely volatile and should be handled with care.

3. Methods of Preparation

Bicyclo[2.2.1]hepten-7-one has been prepared by the oxidation of *anti*-7-hydroxybicyclo[2.2.1]heptene with chromic acid in acetone [2] and with aluminum *t*-butoxide in benzene with benzoquinone as the hydrogen acceptor.[3] The procedure described here is essentially that of Gassman and Pape.[4]

4. Merits of the Preparation

Bicyclo[2.2.1]hepten-7-one is a useful intermediate in the synthesis of a variety of norbornane derivatives. The present procedure involves a four-step synthesis from hexachlorocyclopentadiene with a 39% overall yield. The next best method [3] involves a four-step synthesis from norbornadiene with a 15% overall yield.

1. Department of Chemistry, The Ohio State University, Columbus, Ohio 43210.

2. C. J. Norton, Ph.D. Thesis, Harvard University, 1955.
3. R. K. Bly and R. S. Bly, *J. Org. Chem.*, **28**, 3165 (1963).
4. P. G. Gassman and P. G. Pape, *J. Org. Chem.*, **29**, 160 (1964).

p-BROMODIPHENYLMETHANE

[Methane, (*p*-bromophenyl)phenyl-]

$$C_6H_5CHO + CHCl_3 \xrightarrow{KOH} C_6H_5\underset{\displaystyle CCl_3}{\underset{|}{C}}HOH \xrightarrow[H_2SO_4]{C_6H_5Br}$$

$$p\text{-}BrC_6H_4\underset{\displaystyle CCl_3}{\underset{|}{C}}HC_6H_5 \xrightarrow{KOH} p\text{-}BrC_6H_4CH_2C_6H_5$$

Submitted by A. B. GALUN [1] and A. KALIR [2]
Checked by R. BRESLOW and H. T. BOZIMO

1. Procedure

A. *1-Phenyl-2,2,2-trichloroethanol.* In a 1-l. round-bottomed-flask fitted with a mechanical stirrer, a thermometer, and a powder funnel is placed a solution of 212 g. (2.00 moles) of freshly distilled benzaldehyde in 400 g. (270 ml., 3.35 moles) of chloroform. The mixture is cooled in an ice bath, and 123 g. of commercial powdered potassium hydroxide is added with stirring at such a rate that the temperature of the solution does not exceed 45° (1–1.5 hours). The reaction mixture is stirred and kept at 40–50° for an additional hour and then poured into a solution of 60 ml. of sulfuric acid in 3 l. of water. The resulting two-phase mixture is transferred to a separatory funnel and extracted with three 250-ml. portions of chloroform (a small amount of insoluble, black resinous material is discarded). The combined organic layers are washed with three 100-ml. portions of aqueous 10% sodium carbonate, dried over anhydrous magnesium sulfate, and filtered into a 1-l. flask. The solvent is removed under reduced pressure on a hot water bath. The residue is transferred to a 250-ml. flask and distilled under reduced pressure to give 1-phenyl-2,2,2-trichloroethanol, b.p. 155–165° (26 mm.), 90–100° (0.5 mm.) (Notes 1 and 2). The yield is 170–180 g. (38–40%).

B. *1-p-Bromophenyl-1-phenyl-2,2,2-trichloroethane.* In a 500-

ml. round-bottomed flask fitted with a mechanical stirrer, a dropping funnel, and a thermometer are placed 136 g. (0.60 mole) of 1-phenyl-2,2,2-trichloroethanol and 120 g. (81 ml., 0.77 mole) of bromobenzene. The flask is cooled in an ice-water bath, and a mixture of 120 ml. of concentrated sulfuric acid and 50 ml. of oleum (20% SO_3) is added with stirring at such a rate that the temperature of the reaction mixture does not exceed 10° (*ca.* 45 minutes) (Note 3). The mixture is stirred for another 30 minutes at 10° and for 4–5 hours at room temperature. It is then poured with manual stirring onto 1 kg. of cracked ice, and the mixture is allowed to stand overnight. The precipitate (Note 4) is filtered, washed with water, and recrystallized from 300 ml. of ethanol (Note 5). The yield is 129–162 g. (59–74%), m.p. 95–96° (Note 6).

C. *p-Bromodiphenylmethane.* A 2-l. three-necked flask fitted with a distillation condenser, a thermometer, and an efficient mechanical stirrer is charged with 1.1 l. of diethylene glycol (Note 7) and a solution of 190 g. of potassium hydroxide in 100 ml. of water. The mixture is stirred, and water is distilled until the internal temperature reaches 180°. The resulting solution is allowed to cool to 100° or below, and 146 g. (0.40 mole) of 1-*p*-bromophenyl-1-phenyl-2,2,2-trichloroethane (Note 8) is added. The condenser is set for reflux, and the mixture is stirred and heated to boiling for 5 hours (Note 9). The hot solution is then poured onto 3 kg. of cracked ice, and the mixture is allowed to stand overnight. The oily layer is separated and dissolved in ether (any insoluble material is discarded), and the aqueous layer is extracted with 250 ml. of ether. The combined ethereal solution and extracts are dried over calcium chloride and filtered. The ether is removed under reduced pressure on a hot water bath. The product is distilled under reduced pressure; b.p. 120–130° (3 mm.), 155–163° (13 mm.) (Note 10), n^{24}_{D} 1.6028, d^{24}_{24} 1.342. The yield is 74–79 g. (75–80%) (Note 11).

2. Notes

1. The purpose of the distillation is to separate the product from tars. Therefore no fractionation is required, and the distillation may be carried out rapidly.

2. The carbinol, which has a tendency to supercool, may crystallize overnight; m.p. 38°.

3. Solid material is sometimes deposited on the walls of the reaction flask.

4. In some cases the organic layer separates as an oil; it is then obtained in crystalline form by trituration with 200 ml. of cold methanol, which dissolves the excess of bromobenzene.

5. Wet material may require larger amounts of ethanol.

6. Trituration of the crude precipitate with methanol gives a 90% yield of material, m.p. 90–93°.

7. Eastman Organic Chemicals white label 2,2-oxydiethanol was used.

8. The material should be thoroughly freed of alcohol, preferably over phosphorus pentoxide under reduced pressure, before use. Even traces of alcohol may reduce the yield to 60%.

9. The temperature of the refluxing solution should be above 165°. Efficient stirring is essential; otherwise the precipitating potassium carbonate entrains much material, causing reduction of yield.

10. Good fractionation is not required.

11. Runs on a fourfold scale give the same yield.

3. Methods of Preparation

The procedure for the preparation of 1-phenyl-2,2,2-trichloroethanol is based on the work of Bergmann, Ginsburg, and Lavie.[3] 1-Phenyl-2,2,2-trichloroethanol has also been prepared from phenylmagnesium bromide and chloral.[4]

p-Bromodiphenylmethane has been reported as a product of the reduction of *p*-bromobenzophenone with hydriodic acid and red phosphorus in a sealed tube at 160°.[5] The present method is a modification of the synthesis published by Galun, Kaluszyner, and Bergmann.[6]

4. Merits of the Preparation

In this method inexpensive, commercially available chemicals are used as starting materials. The operations are simple, the yields acceptable, and the final products are free of isomers.

This procedure is especially suited for preparing variously substituted diarylmethanes.[6] The 1,1-diaryl-2,2,2-trichloroethanes may be converted to the corresponding benzophenones via the 1,1-diaryl-2,2-dichloroethylenes[7] and to 1,1-diarylacetic acids.[8]

1. "Zion" Chemical Products, Ltd., Yavne, Israel.
2. Israel Institute for Biological Research, Ness-Ziona, Israel.
3. E. D. Bergmann, D. Ginsburg, and D. Lavie, *J. Am. Chem. Soc.*, **72**, 5012 (1950).
4. P. Hébert, *Bull. Soc. Chim. France*, [4] **27**, 45 (1920); Z. I. Iotsich, *Zh. Russ. Fiz.-Khim. Obshch.*, **34**, 96 (1902).
5. J. H. Speer and A. J. Hill, *J. Org. Chem.*, **2**, 139 (1937).
6. A. B. Galun, A. Kaluszyner, and E. D. Bergmann, *J. Org. Chem.*, **27**, 1426 (1962).
7. O. Grummitt, A. Buck, and A. Jenkins, *J. Am. Chem. Soc.*, **67**, 155 (1945).
8. O. Grummitt, A. Buck, and R. Egan, *Org. Syntheses*, Coll. Vol. **3**, 270 (1955).

3-BROMOPYRENE

(Pyrene, 1-bromo-)

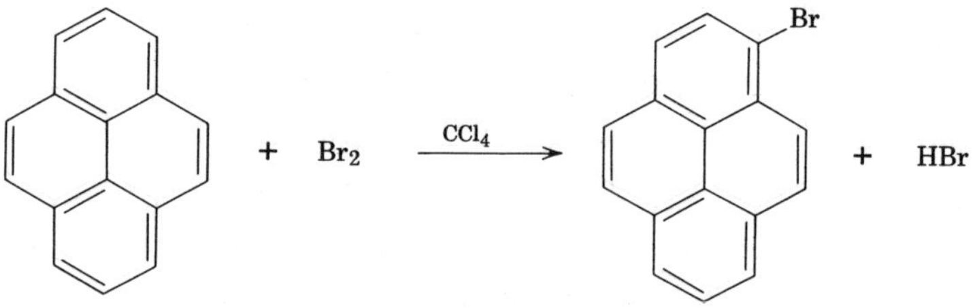

Submitted by W. H. Gumprecht[1]
Checked by Melvin S. Newman and Stephen Havlicek

1. Procedure

Caution! Many pyrenes are carcinogens. Contact of the skin with these materials should be avoided.

In a 500-ml., three-necked, round-bottomed flask fitted with a stirrer, a reflux condenser, and a dropping funnel are placed 8.08 g. (0.040 mole) of pyrene (Note 1) and 80 ml. of carbon tetrachloride (Note 2). A solution of 2.0 ml. of bromine (6.24 g., 0.039 mole) (Note 3) in 30 ml. of carbon tetrachloride is added dropwise over a period of 2–3 hours. The resulting orange solution is stirred overnight, washed with three 100-ml. portions of water, and dried over anhydrous calcium chloride. The solvent is removed under reduced pressure, the pale yellow solid residue is dissolved in 10 ml. of benzene, and the benzene solution is treated with a small amount of activated carbon. The filtrate is diluted with 120 ml. of absolute ethanol, and the solution is distilled until about 80–90 ml. of solvent remains and then cooled. The bromopyrene crystallizes as pale yellow flakes, m.p. 93–95°. Additional material of similar melting point is obtained from the mother liquor on concentration. The total yield is 8.5–9.5 g. (78–86%) (Note 4). On recrystallization from benzene-alcohol a colorless product, m.p. 94.5–95.5°, is obtained with little loss.

2. Notes

1. Pyrene, m.p. 151–153°, obtained from Chemicals Division, Union Carbide Chemicals Corp., was used by the submitter. A commercial pyrene obtained from Germany was used by the checkers.

2. A C.P. solvent was used.

3. Reagent grade material was used. Excess bromine is to be avoided, as dibromopyrene can be formed.[2]

4. The submitter has run this preparation on 80 g. of pyrene with no change in yield.

3. Methods of Preparation

This procedure is described by Lock;[2] a modification using a small amount of phenol has been published.[3] The patent literature discloses the use of a tertiary amine, such as pyridine, and its combination with other solvents for the monobromination of pyrene with elemental bromine.[4] Brominating agents, such as N-bromosuccinimide[5] and N-bromohydantoins,[6] have also been used.

4. Merits of the Preparation

3-Bromopyrene is a precursor of 3-hydroxypyrene.[7]

1. Contribution No. 334 from the Organic Chemicals Department, E. I. du Pont de Nemours and Company, Wilmington, Delaware 12899.
2. G. Lock, *Ber.*, **70**, 926 (1937).
3. H. Hock and F. Ernst, *Ber.*, **92**, 2732 (1959).
4. A. Wolfram and W. Schnurr, U.S. Patent 2,094,227 (1937) [*C.A.*, **31**, 8550 (1937)].
5. Ng. Ph. Buu-Hoi and J. Lecocq, *Compt. Rend.*, **226**, 87 (1948).
6. J. F. Salellas and O. O. Orazi, *Anales Asoc. Quim. Arg.*, **39**, 175 (1951) [*C.A.*, **47**, 2708 (1953)]; R. A. Corral, O. O. Orazi, and J. D. Bonafede, *Anales Asoc. Quim. Arg.*, **45**, 151 (1957) [*C.A.*, **53**, 342 (1959)].
7. W. H. Gumprecht, *Org. Syntheses*, this volume, p. 94.

t-BUTYL CARBAMATE

(Carbamic acid, *tert*-butyl ester)

$$(CH_3)_3COH \xrightarrow[CF_3COOH]{NaCNO} (CH_3)_3CO\overset{\overset{O}{\|}}{C}NH_2$$

Submitted by Bernard Loev, Minerva F. Kormendy, and Marjorie M. Goodman[1]
Checked by David C. Armbruster and William D. Emmons

1. Procedure

Caution! Because of the acrid nature of trifluoroacetic acid and the possibility of the evolution of toxic fumes the reaction should be carried out in a hood.

A solution of 14.8 g. (0.20 mole) of *t*-butyl alcohol in 125 ml. of benzene (Note 1) is placed in a 500-ml. three-necked flask equipped with a stirrer, a thermometer, and an addition funnel, and 26.0 g. (0.40 mole) of sodium cyanate (Note 2) is added. The suspension is stirred as *slowly* as possible (*ca.* 120 r.p.m.; Note 3) while 48.0 g. (31.2 ml., 0.42 mole) of trifluoroacetic acid is added dropwise at a rapid rate. The temperature slowly rises to about 37° after three-quarters of the trifluoroacetic acid has been added (*ca.* 7 minutes). At this point (Note 4) the mixture is cooled to 33–35° by brief immersion in an ice-water bath, then the addition is continued. When the addition of the acid is completed (10–12 minutes total time), the temperature slowly rises to 40° and then gradually subsides. Slow stirring is continued overnight (Note 5) at room temperature.

The mixture is treated with 35 ml. of water (Note 6) and stirred vigorously for a few minutes. The benzene layer is decanted, and the aqueous slurry is rinsed with two 125-ml. portions of benzene (Note 7). The combined organic extracts are washed once with 100 ml. of aqueous 5% sodium hydroxide (Note 8) and with 100 ml. of water, dried over anhydrous magnesium sulfate, and filtered. The solvent is removed by distillation under reduced pressure, preferably on a rotary evaporator, from a water bath kept at 30° (Note 9) to give 17.7–22.0 g. (76–94%) of *t*-butyl carbamate as white needles, m.p. 104–109° (Note 10). The product may be recrystallized from hexane (Note 11); m.p. 107–109° (Note 12).

2. Notes

1. The reagents should not be dried, as traces of moisture catalyze the reaction. The choice of solvent for this type of reaction markedly affects the yield; for most alcohols the use of benzene or methylene chloride gives yields superior to those obtained in other solvents.

2. Sodium cyanate *cannot* be replaced by other cyanates (potassium, ammonium, etc.), for the yields are then drastically lowered.

3. Vigorous agitation markedly lowers the yield; stirring rates of 40–120 r.p.m. are optimum.

4. The temperature may rise to 40°; within the range 20–50° the temperature has little effect on the yield.

5. A contact time of 3–4 hours is sufficient, but it is convenient to stir the reaction mixture overnight. The yield is slightly higher after this additional time.

6. Only a limited amount of water is added at this point because *t*-butyl carbamate has some solubility in the resulting aqueous slurry. With water-insoluble carbamates the amount of water added is immaterial.

7. The checkers found that quantitative recovery of the benzene layer by decantation was impossible, so that in the final

benzene rinse the mixture was poured into a graduated cylinder, and the benzene layer was quantitatively removed by a syringe.

8. The alkaline wash serves to hydrolyze a small amount of *t*-butyl N-trifluoroacetylcarbamate which occasionally forms. It is not clear why this by-product forms on some occasions but not on others under apparently identical conditions. The checkers found in every case that upon standing the alkaline wash deposited 1–2 g. (after drying) of white crystals which was shown to be identical with the *t*-butyl carbamate obtained as the main crop. This amount is included in the yield.

9. Most carbamates, including those of high molecular weight, are volatile. They are generally thermally unstable until they are purified.

10. The melting range varies markedly with the rate of heating, the temperature at which the sample is put into the bath, the solvent used, and the crystal form of the product. The compound at this stage is analytically pure and gives a single spot on thin-layer chromatography.

11. The carbamate may also be recrystallized from water in somewhat lower recovery. With either solvent, extensive heating should be avoided since a considerable amount of product is lost by volatilization. The checkers found that a relatively large volume of hexane was required for recrystallization and therefore used a 1:1 benzene-hexane or 1:1 benzene-ligroin solvent system for the recrystallization.

12. The reported melting points range from 108° to 110°.[2–5]

3. Methods of Preparation

Although numerous methods are known for the synthesis of carbamates of primary and secondary alcohols,[6] they are not satisfactory for the preparation of carbamates of tertiary alcohols.[7, 8] *t*-Butyl carbamate was first obtained by reaction of sodium *t*-butoxide with phosgene and thionyl chloride at −60°, followed by reaction with concentrated aqueous ammonia; the overall yield was less than 20%.[2] This procedure, however, was found to be unsuitable for the preparation of carbamates of other tertiary alcohols.[8] Carbamates have been prepared by the reaction of phenyl chloroformate (prepared from phenol and phosgene at −60°) with a tertiary alcohol in pyridine, followed by treatment with liquid ammonia.[8] A variation of this procedure involves hydrazinolysis of phenyl *t*-butyl carbonate, prepared as described above, conversion to the azide, and ammonolysis.[3, 4] *t*-Butyl carbamate has also been prepared by a four-step procedure that starts with the preparation of *t*-butyl ethyl oxalate from ethoxalyl chloride. This mixed ester was converted to *t*-butyl oxamate, which was dehydrated to *t*-butyl cyanoformate, and this was treated with ammonia.[4]

The carbamates of tertiary acetylenic alcohols have also been made by reaction of these alcohols with sodium cyanate in trifluoroacetic acid.[9] The yields by this procedure are significantly lower than those obtained by the present modification, which is essentially that described by Loev and Kormendy.[5]

4. Merits of the Preparation

This one-step procedure is a convenient and general method for the preparation of carbamates. It is substantially simpler, quicker, and safer than the multistep methods hitherto used for the preparation of carbamates of tertiary alcohols. This procedure is applicable to the preparation of carbamates of primary, secondary, and tertiary alcohols and mercaptans, polyhydric alcohols, acetylenic alcohols, phenols, and oximes. It has also been extended to the preparation of carbamyl derivatives (*i.e.*, ureas) of inert (non-basic) amines.[10]

1. Research and Development Division, Smith, Kline, and French Laboratories, Philadelphia, Pennsylvania 19101.

2. A. R. Choppin and J. W. Rogers, *J. Am. Chem. Soc.*, **70**, 2967 (1948).
3. L. A. Carpino, *J. Am. Chem. Soc.*, **79**, 98 (1957); **79**, 4427 (1957).
4. L. A. Carpino, *J. Am. Chem. Soc.*, **82**, 2725 (1960).
5. B. Loev and M. F. Kormendy, *J. Org. Chem.*, **28**, 3421 (1963).
6. S. Petersen, in E. Müller, "Methoden der Organischen Chemie (Houben-Weyl)," 4th ed., Vol. 8, Georg Thieme Verlag, Stuttgart, 1952, p. 137.
7. L. Yoder, *J. Am. Chem. Soc.*, **45**, 475 (1923).
8. W. M. McLamore, S. Y. P'an, and A. Bavley, *J. Org. Chem.*, **20**, 1379 (1955).
9. P. G. Marshall, J. H. Barnes, and P. A. McCrea, U. S. Patent 2,814,637 (1957) [*C.A.*, **52**, 2901 (1958)].
10. G. J. Durant, *Chem. Ind.* (*London*), 1428 (1965).

t-BUTYL DIAZOACETATE

(Acetic acid, diazo-, *tert*-butyl ester)

$$p\text{-}CH_3C_6H_4SO_2Cl + NaN_3 \rightarrow p\text{-}CH_3C_6H_4SO_2N_3$$

$$CH_3COCH_2CO_2C(CH_3)_3 + p\text{-}CH_3C_6H_4SO_2N_3 \xrightarrow{(C_2H_5)_3N} CH_3COCN_2CO_2C(CH_3)_3 + p\text{-}CH_3C_6H_4SO_2NH_2$$

$$CH_3COCN_2CO_2C(CH_3)_3 \xrightarrow[CH_3OH]{CH_3ONa} N_2CHCO_2C(CH_3)_3 + CH_3CO_2CH_3$$

Submitted by Manfred Regitz, Jürgen Hocker, and Annemarie Liedhegener [1]
Checked by C. John Blankley and Herbert O. House

1. Procedure

Caution! Diazoacetic esters are toxic and potentially explosive and must be handled with caution. This preparation should be carried out in a hood, and the distillation of t-butyl diazoacetate should be conducted behind a safety shield.

A. *p-Toluenesulfonyl azide.*[2, 3] A solution of 71.5 g. (1.10 moles) of sodium azide (Note 1) in 200 ml. of water is placed in a 2-l. Erlenmeyer flask and diluted with 400 ml. of 90% aqueous ethanol (Note 2). To this solution is added with stirring a warm (45°) solution of 190.5 g. (1.00 mole) of *p*-toluenesulfonyl chloride (Note 3) in 1 l. of 99% ethanol (Note 2). During this addition, sodium chloride separates, and the reaction mixture takes on a light brown color. After the reaction mixture has been stirred at room temperature for 2.5 hours, most of the solvent is removed at 35° (15 mm.) with a rotary evaporator (Note 4). The residue is mixed with 1.2 l. of water in a separatory funnel, and the oily *p*-toluenesulfonyl azide is separated. This oil is washed with two 100-ml. portions of water and dried over anhydrous sodium sulfate. Filtration with suction gives 160–170 g. (81–86%, based on *p*-toluenesulfonyl chloride) of pure, colorless *p*-toluenesulfonyl azide which completely crystallizes on standing at 5°.

B. *t-Butyl α-diazoacetoacetate.* In a 2-l., wide-mouthed, Erlenmeyer flask are placed 118.5 g. (0.75 mole) of *t*-butyl acetoacetate (Note 5), 1 l. of anhydrous acetonitrile, and 75.8 g. (0.75 mole) of previously distilled triethylamine (b.p. 88.5–90.5°). The temperature of the mixture is adjusted to 20°, and 148 g. (0.75 mole) of *p*-toluenesulfonyl azide is added dropwise with vigorous stirring over 10–15 minutes. The addition causes the reaction mixture to warm to 38–40° and assume a yellow color. After the mixture has been stirred at room temperature for 2.5 hours, the solvent is evaporated at 35° (12 mm.). The partially crystalline residue is triturated with 1 l. of ether, and the mixture, including the insoluble residue, is placed in a 2-l. separatory funnel. The mixture is washed successively with a solution of 45 g. of potassium hydroxide in 500 ml. of water, a solution of 7.5 g. of potassium hydroxide in 250 ml. of water, and 250 ml. of water (Note 6). The yellow-orange ethereal phase is dried over anhydrous sodium sulfate, and the solvent is evaporated at 35° (15 mm.) until the residue has attained a constant weight. The yellow-orange diazo ester weighs 130–135 g. (94–98%) (Note 7).

C. *t-Butyl diazoacetate.* Into a 1-l. three-necked flask fitted with a stirrer, a dropping funnel, and a thermometer is placed a solution of 92.6 (0.50 mole) of *t*-butyl α-diazoacetoacetate in 150 ml. of methanol. After this solution has been cooled to 2–3° in an ice bath, a solution of sodium methoxide, prepared from 11.5 g. (0.50 g. atom) of sodium and 150 ml. of methanol, is added dropwise with stirring at such a rate that the reaction mixture remains within the temperature range 0–5° (about 30 minutes is required for the addition). After the addition is completed, the mixture is stirred in the ice bath for an additional 30 minutes. The red reaction solution is poured into 1 l. of ice water, and the resulting mixture is extracted with 500 ml. of ether. The aqueous phase is saturated with sodium chloride and extracted with two 500-ml. portions of ether (Note 8). The combined ethereal extracts are washed with 500 ml. of water and dried over anhydrous sodium sulfate. After the mixture has been filtered and the residue has been washed with ether, the bulk of the solvent is removed from the combined ethereal filtrates at 30° and water aspirator pressure with a rotary evaporator (Note 9). The remaining ether is removed by distillation under slightly reduced pressure while the stillpot is heated with a water bath at 50°. The residual red oil is distilled. (*Caution! See above.*) (Note 10). After a small forerun the diazo ester distills during which time the temperature of the water bath is raised from 60° to 75°. The yield is 48–50 g. (68–70%) of yellow-orange liquid, b.p. 51–53° (12 mm.), n^{20}D 1.4551, $R_f = 0.56$ (chloroform) (Note 11).

2. Notes

1. The submitters used sodium azide obtained from Dr. F. Raschig, GmbH, 67 Ludwigshafen, Rhein, Germany. The checkers used material from Eastman Organic Chemicals.

2. The checkers found 95% ethanol denatured with methanol to be a satisfactory substitute.

3. The submitters used *p*-toluenesulfonyl chloride obtained from Badische Anilin- und Soda-Fabrik, 67 Ludwigshafen, Rhein, Germany. Very impure *p*-toluenesulfonyl chloride can be purified by recrystallization from ether. The checkers used material from Matheson, Coleman and Bell without further purification.

4. In order to prevent foaming, the concentration is begun with the water bath at *ca.* 10°, and the bath is warmed slowly to 35°.

5. *t*-Butyl acetoacetate may be prepared from *t*-butyl alcohol and diketene.[4] The checkers obtained this material from Eastman Chemical Products, Inc.

6. Acidification of the aqueous potassium hydroxide phase with 6*N* hydrochloric acid gives *p*-toluenesulfonamide. After being dried at 85° (50 mm.) the sample weighs 110–120 g. (86–94%) and melts at 132–134°.

7. If desired, the α-diazo β-keto ester can be purified by a low-temperature crystallization. The diazo ester (10 g.) is cooled to −70° to −75° in a dry ice-acetone bath, and crystallization is initiated by rubbing. (*Caution! The rubbing should not be continued after crystallization has been initiated.*) This material is treated with 5 ml. of anhydrous ether which has been previously cooled, and the mixture is filtered with suction. The residue

from the filtration is placed in a flask, and the residual ether is removed by evaporation at 35° (15 mm.) to give 5–6 g. of the yellow diazo ester.[5]

8. If the ethereal phase contains a small amount of insoluble material, the mixture should be filtered to avoid difficulty in separating the phases.

9. The distillate is light yellow and contains some *t*-butyl diazoacetate.

10. This distillation has been conducted with the usual precautions (safety glasses, safety shield) with no explosions up to the present time.

11. The thin-layer chromatogram was obtained on "DC-Fertiplatte Merck Kieselgel F_{254}" purchased from E. Merck A. G., 61 Darmstadt, Germany. Employing an Eastman Chromatoplate K301R1 (silica without indicator) with chloroform as eluent, the checkers found an R_f value of 0.72.

3. Methods of Preparation

t-Butyl diazoacetate has been prepared by the present method, by alkaline decomposition of *t*-butyl N-nitroso-N-acetylglycinate,[6] and by diazotization of *t*-butyl glycinate.[7]

4. Merits of the Preparation

The transformation of an active CH compound into the corresponding diazo derivative with *p*-toluenesulfonyl azide has been designated a "diazo transfer reaction"[8] and possesses a variety of preparative uses. The method has been useful for the syntheses of diazo derivatives of cyclopentadiene,[3, 9] 1,3-dicarbonyl compounds,[5, 10, 11] 1,3-disulfonyl compounds,[12] 1,3-ketosulfonyl compounds,[13, 14] ketones,[15-17] carboxylic acid esters,[15, 17] and β-keto imines.[18] Further reaction of these diazo intermediates can lead to azo compounds,[11, 13] 1,2,3-triazoles,[18, 19] and pyrazolinones.[8]

1. Institut für Organische Chemie der Universität des Saarlandes, 66 Saarbrücken, Germany.
2. T. Curtius and G. Kraemer, *J. Prakt. Chem.*, [2] **125**, 303 (1930).

3. W. von E. Doering and C. H. DePuy, *J. Am. Chem. Soc.*, **75**, 5955 (1953).
4. S.-O. Lawesson, S. Gronwall, and R. Sandberg, *Org. Syntheses*, **42**, 28 (1962).
5. M. Regitz and A. Liedhegener, *Ber.*, **99**, 3128 (1966).
6. H. Reimlinger and L. Skatteböl, *Ber.*, **93**, 2162 (1960).
7. E. Müller and H. Huber-Emden, *Ann.*, **660**, 54 (1962).
8. M. Regitz, *Angew. Chem.*, **79**, 786 (1967); *Angew. Chem. Intern. Ed. Engl.*, **6**, 733 (1967).
9. M. Regitz and A. Liedhegener, *Tetrahedron*, **23**, 2701 (1967).
10. M. Regitz, *Ann.*, **676**, 101 (1964).
11. M. Regitz and D. Stadler, *Ann.*, **687**, 214 (1965).
12. F. Klages and K. Bott, *Ber.*, **97**, 735 (1964).
13. M. Regitz, *Ber.*, **98**, 36 (1965).
14. A. M. van Leusen, P. M. Smid, and J. Strating, *Tetrahedron Letters*, 337 (1965).
15. M. Regitz, *Ber.*, **98**, 1210 (1965); *Tetrahedron Letters*, 1403 (1964).
16. M. Rosenberger, P. Yates, J. B. Hendrickson, and W. Wolf, *Tetrahedron Letters*, 2285 (1964).
17. M. Regitz, F. Menz, and J. Rüter, *Tetrahedron Letters*, 739 (1967).
18. M. Regitz, *Tetrahedron Letters*, 3287 (1965).
19. M. Regitz and A. Liedhegener, *Ber.*, **99**, 2918 (1966).

2-*n*-BUTYL-2-METHYLCYCLOHEXANONE

(Cyclohexanone, 2-butyl-2-methyl-)

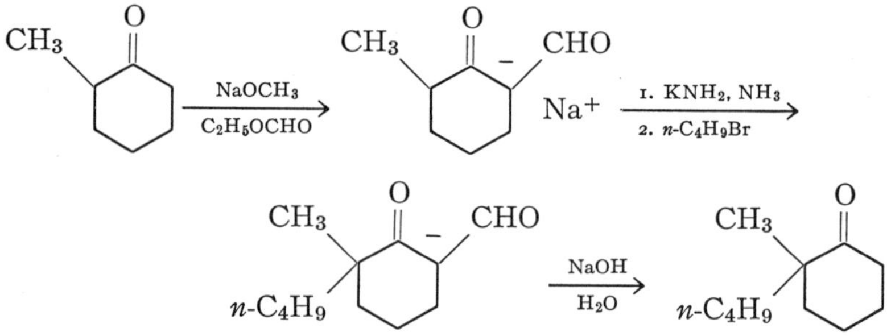

Submitted by S. Boatman, T. M. Harris, and C. R. Hauser [1]
Checked by William G. Dauben, Michael H. McGann, and Noel Vietmeyer

1. Procedure

Caution! This preparation should be carried out in a hood to avoid exposure to ammonia.

In a 3-l. three-necked flask fitted with a calcium chloride drying tube, a nitrogen-inlet tube, and a sealed mechanical stirrer are placed 54.0 g. (1.00 mole) of commercial, anhydrous sodium methoxide (Note 1) and 2 l. of anhydrous ether. The flask is purged with dry nitrogen and cooled in an ice bath. The inlet tube is replaced by an addition funnel containing a solution of 123 g. (1.10 moles) of 2-methylcyclohexanone (Note 2) and 81.4 g. (1.10 moles) of ethyl formate (Note 3). The solution is added rapidly, dropwise, and at the end of the addition the funnel is replaced by the nitrogen-inlet tube. After 15 minutes the ice bath is removed, and the mixture is stirred for 12 hours at room temperature. The thick suspension is filtered by suction, and the filter cake is washed with anhydrous ether, care being taken to protect the product from atmospheric moisture (Note 4). The solid salt is dried in a vacuum oven at *ca.* 70°, powdered (Note 5), and stored in a tightly capped bottle. Sodio-2-formyl-6-methylcyclohexanone, a cream-colored powder, is obtained in 80–85% (130–138 g.) yield (Note 6).

In a 1-l. three-necked flask equipped with a dry ice-acetone condenser and a sealed mechanical stirrer is placed 700 ml. of commercial, anhydrous, liquid ammonia. To the stirred ammonia is added a small piece of potassium metal. (*Caution! Care should be exercised in handling potassium metal, since it is extremely reactive and it ignites on contact with water, atmospheric moisture, or alcohol. It should be manipulated under toluene or xylene, and blotted with filter paper before addition.*) After the appearance of a blue color a few crystals of ferric nitrate hydrate (*ca.* 0.1 g.) are added, followed by small pieces of freshly cut potassium metal until 7.0 g. (0.18 g. atom) has been added. After all the potassium has been converted to the amide (Note 7), 24.9 g. (0.154 mole) of sodio-2-formyl-6-methylcyclohexanone is added carefully through a powder funnel (Note 8). After 1 hour a solution of 28.2 g. (0.21 mole) of *n*-butyl bromide (Note 9) in 50 ml. of anhydrous ether is added dropwise from an addition funnel. The mixture is stirred for 3 hours, and then the dry ice-acetone condenser is replaced by a water condenser. A steam bath is placed under the flask, and the ammonia is evaporated (*Caution!*) as 400 ml. of anhydrous ether is added. When the ammonia has been removed and the ether has refluxed for 5 minutes, 100 g. of ice is added, followed by 300 ml. of water. When the solid has dissolved, the layers are separated, and the ethereal layer is extracted twice with cold water. The combined aqueous extracts are placed in a 1-l. round-bottomed flask, and 6.4 g. of sodium hydroxide is added. The flask is warmed briefly to remove dissolved ether from the solution. The flask is equipped with an efficient condenser, and the mixture is refluxed until an enol test is no longer obtained (6–8 hours) (Notes 10 and 11). The mixture is cooled and extracted with three 200-ml. portions of ether. The combined ethereal extracts are washed with dilute hydrochloric acid and dried over anhydrous magnesium sulfate. The ether is evaporated, and the residue is distilled under reduced pressure to give 14–19 g. (54–74%) of 2-*n*-butyl-2-methylcyclohexanone, b.p. 116–118° (20 mm.) (Note 12).

2. Notes

1. Sodium methoxide was obtained from Matheson, Coleman and Bell. The best results were obtained with material from freshly opened bottles.

2. Eastman Organic Chemicals "Eastman grade" 2-methylcyclohexanone was distilled; b.p. 56° (20 mm.).

3. Eastman Organic Chemicals practical grade ethyl formate was shaken for 30 minutes with anhydrous sodium carbonate and for 30 minutes with anhydrous magnesium sulfate, and distilled; b.p. 54°.

4. A rubber dam was fastened tightly over the top of the Buchner funnel by means of rubber bands. It was pulled down onto the surface of the filter cake by the vacuum.

5. The solid should be powdered to allow complete formation of the dianion in the following reaction. This is most readily accomplished if the solid is ground before it is completely dry (*i.e.*, when it appears to be dry but is still cool). The fine powder is then replaced in the oven to complete the drying.

6. The checkers, working at one-quarter scale, obtained a yield of 86–88%.

7. Conversion is indicated by discharge of the deep blue color.

This generally requires about 20 minutes. When conversion is completed, the stirrer should be speeded up or the contents of the flask swirled so that potassium splattered on the upper part of the flask is converted to amide; this should be done until all traces of blue color are gone.

8. The escaping ammonia will blow away some of the fine powder unless this is done carefully.

9. Eastman Organic Chemicals "Eastman grade" *n*-butyl bromide was distilled; b.p. 101–102°.

10. The enol test is performed with about 0.5 ml. of solution, which is neutralized with dilute hydrochloric acid and treated with 3–5 drops of 10% ethanolic ferric chloride. A reddish brown color denotes the presence of unhydrolyzed formyl ketone.

11. An alternative procedure is steam distillation of the basic, aqueous solution until no further organic material distills. This may be done either instead of, or after, the refluxing of the aqueous solution. The steam distillate is extracted with ether, and the ether is removed by distillation.

12. A higher-boiling fraction consisting of 2-formyl-6-*n*-butyl-6-methylcyclohexanone, b.p. 201–203° (20 mm.), is obtained if hydrolysis is not complete.

3. Methods of Preparation

This procedure is an adaptation of one described by Boatman, Harris, and Hauser.[2]

4. Merits of the Preparation

The present method affords 2-*n*-butyl-2-methylcyclohexanone uncontaminated by the isomeric 2-*n*-butyl-6-methylcyclohexanone.

2,2-Dimethylcyclohexanone and 2-benzyl-2-methylcyclohexanone have been prepared similarly in yields of 60% and 55%, respectively.[2] The procedure has been extended to the synthesis of 9-methyl-, 9-*n*-butyl-, and 9-benzyl-1-decalone from the dianion of 2-formyl-1-decalone in yields of 55%, 48%, and 58%, respectively.[2]

1. Department of Chemistry, Duke University, Durham, North Carolina 27706.
2. S. Boatman, T. M. Harris, and C. R. Hauser, *J. Am. Chem. Soc.*, **87**, 82 (1965).

1,1′-CARBONYLDIIMIDAZOLE

(Imidazole, 1,1′-carbonyldi-)

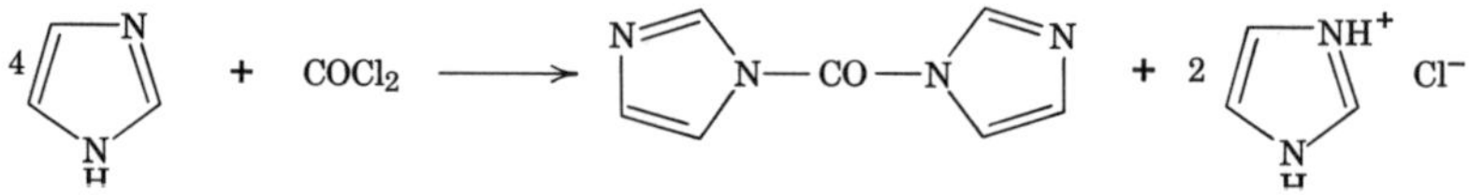

Submitted by HEINZ A. STAAB and KURT WENDEL[1]
Checked by A. C. MACKAY and PETER YATES

1. Procedure

Caution! This preparation must be carried out in a hood to avoid exposure to phosgene.

Anhydrous benzene (*ca.* 200 ml.) (Note 1) is poured into a calibrated, 500-ml., standard-taper dropping funnel equipped with a gas-inlet tube containing a fritted-glass filter; the dropping funnel is stoppered and weighed accurately. The funnel is protected with a calcium chloride tube, and 15–20 g. of phosgene is introduced at room temperature over a period of *ca.* 1 hour; this quantity corresponds to an increase in volume of 12–16 ml. (Note 2). The calcium chloride tube is removed, and the funnel is immediately restoppered and reweighed (Note 3). The amount of imidazole corresponding to the increase in weight observed (*e.g.*, 16.55 g., 0.167 mole, of phosgene) is calculated on the basis of a phosgene:imidazole molar ratio of 1:4 (Note 4). The funnel is placed on a 1-l., three-necked, round-bottomed flask, that contains a solution of the imidazole (here, 45.60 g., 0.669 mole) in 500 ml. of anhydrous tetrahydrofuran (Note 5) and is equipped with a sealed mechanical stirrer and a calcium chloride tube. The flask is cooled with cold water, and the solution of phosgene in benzene is added with stirring from the dropping funnel over a period of 15–30 minutes. The reaction mixture is stirred for an additional 15 minutes and then allowed to stand for 1 hour at room temperature. The precipitate of imidazolium chloride is removed by suction filtration with exclusion of atmospheric moisture by the use of a standard-taper fritted-glass filter funnel (Note 6). The filtrate is evaporated to dryness under reduced pressure on a water bath at 40–50° (Note 7). The yield of colorless crystalline 1,1′-carbonyldiimidazole is 80–94% (here, 24.8 g.; 91%). The product obtained in this way sinters at 110° and melts between 112° and 117° (here, 114–115°). This material can be used without further purification for most reactions, *e.g.*, ester, peptide, and aldehyde syntheses [2] (Note 8). The purity of a product of m.p. 113–117° was 98±2%.[3] The quality and yield of 1,1′-carbonyldiimidazole are not reduced when the scale is doubled.

1,1′-Carbonyldiimidazole may be kept for a long period of time in either a desiccator over phosphorus pentoxide or in a sealed tube. It is hydrolyzed by water to give carbon dioxide and imidazole.

2. Notes

1. The benzene is heated under reflux over sodium with benzophenone until a permanent blue coloration develops and then is distilled with exclusion of atmospheric moisture.

2. Use of a calibrated dropping funnel permits approximate estimation of the amount of phosgene absorbed; a volume increase of 1 ml. corresponds to about 1.3 g. of phosgene. The stream of phosgene is led through a wash bottle containing concentrated sulfuric acid and should not be too fast in order to avoid loss of solvent by evaporation.

3. The checkers used a fritted-glass inlet tube and drying tube incorporated in a standard-taper adapter that fitted the neck of the dropping funnel. Owing to small losses of solution on withdrawal of the adapter, the weight of phosgene was slightly underestimated.

4. Technical grade imidazole (from Badische Anilin- und Soda-Fabrik, Ludwigshafen, Rhein, Germany) was recrystallized from benzene containing 1.0–1.5% ethanol; m.p. 90°. The checkers used imidazole obtained from Aldrich Chemical Co. without further purification.

5. Technical grade tetrahydrofuran was predried for a few days over sodium hydroxide. It was then heated under reflux over sodium wire with benzophenone until it developed a permanent blue color and distilled with exclusion of atmospheric moisture. [*Caution! See Org. Syntheses*, **46**, *105 (1966), for a warning regarding purification of tetrahydrofuran.*]

6. By working quickly, the imidazolium chloride may be removed by suction filtration through a Buchner funnel. However, the precipitate should not be freed of solvent completely because imidazolium chloride is extremely hygroscopic. If the moist precipitate is washed with 50–100 ml. of anhydrous tetrahydrofuran, the yield of 1,1′-carbonyldiimidazole may be slightly increased; however, there is some danger of the introduction of too much moisture into the reaction solution.

7. The checkers used an antifoaming head for the solvent evaporation.

8. In order to obtain a purer product the crude material may be recrystallized from hot anhydrous tetrahydrofuran with careful exclusion of moisture. After this operation the yield is reduced to 65–75%; the m.p. is then between 114° and 118°; *e.g.*, recrystallization of 24.8 g. (91%) of 1,1′-carbonyldiimidazole from 60 ml. of anhydrous tetrahydrofuran yielded 19.9 g. (73%); m.p. 116–118°.

3. Methods of Preparation

1,1′-Carbonyldiimidazole has been prepared by the reaction of imidazole and phosgene in anhydrous benzene and anhydrous tetrahydrofuran.[3-5] It has also been obtained by the reaction of 1-(trimethylsilyl)imidazole and phosgene in anhydrous benzene,[6] but that method offers no advantages that justify the more extensive preparative effort required.

4. Merits of the Preparation

1,1′-Carbonyldiimidazole has been used for the preparation of such compounds as esters, anhydrides, amides, peptides, ketones, ethers, and isocyanates.[2] The present procedure provides a convenient method for its preparation in good yield.

1. Department of Chemistry, University of Heidelberg, Heidelberg, Germany.
2. H. A. Staab, *Angew. Chem.*, **74**, 407 (1962); *Angew. Chem. Intern. Ed. Engl.*, **1**, 351 (1962).
3. R. Paul and G. W. Anderson, *J. Am. Chem. Soc.*, **82**, 4596 (1960); G. W. Anderson and R. Paul, *J. Am. Chem. Soc.*, **80**, 4423 (1958).
4. H. A. Staab, *Ann.*, **609**, 75 (1957).
5. H. A. Staab and K. Wendel, *Ber.*, **93**, 2910 (1960).
6. L. Birkofer, P. Richter, and A. Ritter, *Ber.*, **93**, 2804 (1960).

CHLORODIISOPROPYLPHOSPHINE

(Phosphinous chloride, diisopropyl-)

$$2(CH_3)_2CHMgCl + PCl_3 \rightarrow [(CH_3)_2CH]_2PCl + 2MgCl_2$$

Submitted by W. Voskuil and J. F. Arens [1]
Checked by Hugh D. Olmstead, James E. Oliver, and Herbert O. House

1. Procedure

Caution! Because of the sensitivity of the reagents and product to moisture and oxygen, all manipulations must be performed in an anhydrous, inert atmosphere (Note 1).

A 500-ml., four-necked, round-bottomed flask is equipped with an efficient stirrer, a reflux condenser, a 250-ml. dropping funnel, and a low-temperature thermometer (Note 2). In the flask are placed 34.4 g. (21.8 ml., 0.25 mole) of phosphorus trichloride (Note 3) and 150 ml. of anhydrous ether. A solution of 0.50 mole of isopropylmagnesium chloride in about 150 ml. of ether (Notes 4 and 5) is placed in the dropping funnel.

The flask is cooled in a dry ice-acetone bath, and the Grignard reagent solution is added dropwise with rapid stirring at such a rate that the temperature of the reaction mixture remains between −25° and −30° with a bath temperature of −45°; this addition requires about 1.5 hours. After the addition has been completed, the cooling bath is removed, and the mixture is allowed to warm to room temperature. Finally, the reaction mixture is heated to reflux with continuous stirring for 30 minutes.

After the reaction mixture has cooled to room temperature, it is filtered with suction (Notes 6 and 7), and the residual salts are washed thoroughly with three 100-ml. portions of anhydrous ether. The combined ethereal filtrates are concentrated under reduced pressure at room temperature, and the residual liquid is fractionally distilled through a 15-cm. Vigreux column. After a small forerun has been collected, the product is obtained as a clear, colorless liquid, b.p. 46–47° (10 mm.), n^{20}_{D} 1.4752 (Note 8). The yield is 21–23 g. (55–60%); practically no residue remains in the distillation pot.

2. Notes

1. The submitters used nitrogen purified by passage through B.T.S. catalyst (B.A.S.F., Ludwigshafen, Germany). The checkers used commercial prepurified nitrogen without further treatment.
2. The checkers used a three-necked flask, one neck of which was fitted with an adapter to accommodate the thermometer. They also used a pressure-equalizing dropping funnel so that a static nitrogen atmosphere several millimeters above atmospheric pressure could be maintained in the flask.
3. The submitters used Merck reagent grade phosphorus trichloride. The checkers used material from Baker and Adamson.
4. It is essential to use the Grignard reagent prepared from isopropyl chloride. From phosphorus trichloride and isopropylmagnesium bromide, bromodiisopropylphosphine is obtained because of a halogen exchange reaction between the initially formed chlorophosphine and magnesium bromide. The checkers used both the Grignard reagent prepared from isopropyl chloride and a commercial solution of isopropylmagnesium chloride available from Matheson, Coleman and Bell.
5. The concentration of the Grignard reagent should be estimated by titration. If an excess or less than the stoichiometric amount of the organometallic reagent is added, the yield is lower and the product is less pure. The checkers found the titration procedure of Watson and Eastham [2] to be most convenient. In a typical titration, performed under a nitrogen atmosphere, a

5.00-ml. aliquot of the Grignard reagent was added to a solution of about 2 mg. of *o*-phenanthroline in 10 ml. of anhydrous benzene. The resulting purple solution was titrated with a standard solution (0.999*M*) of *sec*-butyl alcohol in xylene until the purple color of the *o*-phenanthroline-Grignard reagent charge transfer complex was just discharged. In this procedure the number of millimoles of *sec*-butyl alcohol added is equal to the number of millimoles of alkylmagnesium chloride present in the aliquot of Grignard reagent.

6. The checkers performed this filtration and subsequent washing of the precipitate by replacing the dropping funnel in the reaction flask by a sintered-glass filter stick. A slight positive nitrogen pressure was applied in the reaction flask, and the pressure was reduced in the flask that served as a receiver for the filtrate passing through the sintered-glass filter.

7. The checkers found it necessary to dislodge and break up the cake of magnesium salts that formed on the walls of the reaction flask. If this precaution was not observed, a substantial amount of product occluded in the salt cake was not recovered during the washing process.

8. The checkers verified the absence of dichloroalkylphosphine and trialkylphosphine contaminants in this product by obtaining acceptable elemental analytical results and by measuring the mass spectrum of the product, which exhibits a molecular ion peak at m/e 152 (^{35}Cl) with abundant fragment peaks at m/e 110, 43, and 41.

3. Methods of Preparation

Chlorodiisopropylphosphine has been prepared by the reduction of the diisopropyltrichlorophosphorus-aluminum chloride complex with antimony;[3, 4] this is a general method and the reduction can be performed with other reagents.[5] Other general methods for the preparation of chlorodialkylphosphines are reaction of dialkylphosphines with phosgene [6, 7] and the cleavage of N,N-dialkylaminodialkylphosphines with hydrogen chloride [8-10] or phosphorus trichloride.[11]

4. Merits of the Preparation

Chlorodialkylphosphines are important synthetic intermediates in organophosphorus chemistry. In the chemical literature there is a widespread view that the simple one-step Grignard method is not suitable for the preparation of these compounds because of dominant trisubstitution and the formation of difficultly separable mixtures.[12] Although this is true for the *n*-alkyl compounds, the present preparation demonstrates that in the case of branched primary alkyl compounds and secondary and tertiary alkyl compounds the method can be very convenient and can give pure products. The submitters have prepared[13] chlorodiisobutylphosphine (45–50%), chlorodi-*sec*-butylphosphine (75–80%), chlorodi-*t*-butylphosphine (65–70%), and chlorodicyclohexylphosphine (60–65%) in analogous manner.

With *t*-butylmagnesium chloride the substitution of only one chlorine atom of the phosphorus trichloride is possible, giving dichloro-*t*-butylphosphine (65–70%).

1. Laboratory of Organic Chemistry of the University of Utrecht, Netherlands.
2. S. C. Watson and J. F. Eastham, *J. Organometal. Chem.*, **9**, 165 (1967).
3. J. L. Ferron, *Nature*, **189**, 916 (1961).
4. J. L. Ferron, *Can. J. Chem.*, **39**, 842 (1961).
5. I. P. Komkov, K. V. Karavanov, and S. Z. Ivin, *Zh. Obshch. Khim.*, **28**, 2963 (1958); *J. Gen. Chem. USSR (Engl. Transl.)*, **28**, 2992 (1958).
6. R. Rabinowitz and J. Pellon, *J. Org. Chem.*, **26**, 4623 (1961).
7. W. A. Henderson, Jr., S. A. Buckler, N. E. Day, and M. Grayson, *J. Org. Chem.*, **26**, 4770 (1961).
8. A. B. Burg and P. J. Slota, Jr., *J. Am. Chem. Soc.*, **80**, 1107 (1958).
9. K. Issleib and W. Seidel, *Ber.*, **92**, 2681 (1959).
10. W. Voskuil and J. F. Arens, *Rec. Trav. Chim.*, **81**, 993 (1962).
11. Monsanto Co., Brit. Patent 1,068,364 (1967) [*C.A.*, **67**, 54260 (1967)].
12. Cf. K. Sasse, in E. Müller, "Methoden der Organischen Chemie (Houben-Weyl)," 4th ed., Vol. 12, Part I, Georg Thieme Verlag, Stuttgart, 1963, p. 203.
13. W. Voskuil and J. F. Arens, *Rec. Trav. Chim.*, **82**, 302 (1963).

CINNAMYL BROMIDE

(Benzene, 3-bromopropenyl-)

$$(C_6H_5)_3P + Br_2 \longrightarrow (C_6H_5)_3PBr_2$$

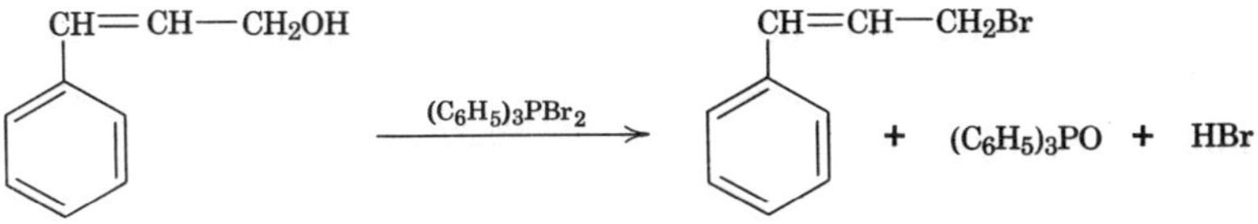

Submitted by JOHN P. SCHAEFER, J. G. HIGGINS, and P. K. SHENOV [1]
Checked by R. BRESLOW and J. T. GROVES

1. Procedure

A 1-l., three-necked, round-bottomed flask equipped with a Trubore stirrer, a pressure-equalizing dropping funnel, and a reflux condenser with a drying tube is charged with 350 ml. of acetonitrile (Note 1) and 106.4 g. (0.41 mole) of triphenylphosphine (Note 2). The flask is cooled in an ice-water bath (Note 3), and 64 g. (0.40 mole) of bromine is added dropwise over a period of *ca.* 15–20 minutes (Notes 4 and 5). The ice-water bath is removed, and a solution of 54 g. (0.40 mole) of cinnamyl alcohol in 50 ml. of acetonitrile is added in portions over a period of 5–10 minutes with continued stirring (Note 6). The solvent is removed by distillation with the use of a water aspirator (30–40 mm.) and an oil bath until the bath temperature reaches 120°. The water aspirator is replaced by a vacuum pump and the water-cooled condenser with an air condenser, and the distillation is continued with rapid stirring (Notes 7, 8, and 9). Most of the product (Note 10) distills at 91–98° (2–4 mm.), and about 59 g. of product crystallizes in the receiving flask (63–75% yield) (Note 11).

The product is dissolved in 200 ml. of ether, and the solution is washed with 75 ml. of saturated aqueous sodium carbonate, dried over anhydrous magnesium sulfate, and distilled to give 47–56 g. (60–71%) of product, b.p. 66–68° (0.07 mm.), 84–86° (0.8 mm.); m.p. 29°.

2. Notes

1. The acetonitrile was distilled from phosphorus pentoxide.
2. Triphenylphosphine was obtained from M and T Chemicals, Inc., and used without further purification.
3. The triphenylphosphine is only partially dissolved.
4. If a slight excess of bromine persists after addition, a small amount of triphenylphosphine should be added until the color of bromine disappears.
5. The solid triphenylphosphine disappears, but at the same time the adduct, $(C_6H_5)_3PBr_2$, precipitates as a white solid.
6. This addition is mildly exothermic, and the temperature rises to 50–60°. All the precipitate should dissolve at this point; warming by external heat may be necessary.
7. To protect the vacuum pump from damage a dry ice-acetone trap and two liquid nitrogen traps are necessary to condense and solidify the hydrogen bromide evolved.
8. The receiving flask is placed in an ice-water bath.
9. The distillation is continued until the triphenylphosphine oxide solidifies and no more product distills. The oil bath is maintained at 130–140° during the distillation.
10. Some product is carried over by the hydrogen bromide in the initial stages of the distillation.
11. When this distillation was replaced by a procedure in which the acetonitrile was removed with a rotary evaporator and steam bath, and the product was extracted from the triphenylphosphine oxide with small portions of acetonitrile totaling *ca.* 250 ml., the checkers obtained an improved yield (79%) of cinnamyl bromide.

3. Methods of Preparation

Cinnamyl bromide has been prepared from cinnamyl alcohol by the action of hydrogen bromide in cold acetic acid [2] and of phosphorus tribromide in boiling benzene.[3] It has also been prepared by the action of N-bromosuccinimide on 3-phenylpropene [4] and on 1-phenylpropene.[5]

4. Merits of the Preparation

The method described is general for converting alcohols to alkyl halides and is stereospecific.[6]

1. Department of Chemistry, The University of Arizona, Tucson, Arizona 85721.
2. J. von Braun and Z. Kohler, *Ber.*, **51**, 79 (1918).
3. M. Gredy, *Bull. Soc. Chim. France*, [5] **3**, 1098 (1936).
4. M. Lora-Tomayo, F. Martin-Panizo, and R. P. Ossorio, *J. Chem. Soc.*, 1418 (1950); E. A. Braude and E. S. Waight, *J. Chem. Soc.*, 1116 (1952).
5. O. A. Orio, *Rev. Fac. Cienc. Quim., Univ. Nacl. La Plata*, **30**, 21 (1957) [*C.A.*, **54**, 12062 (1960)].
6. L. Horner, H. Oediger, and H. Hoffmann, *Ann.*, **626**, 26 (1959); G. A. Wiley, R. L. Hershkowitz, B. M. Rein, and B. C. Chung, *J. Am. Chem. Soc.*, **86**, 964 (1964); J. P. Schaefer and D. S. Weinberg, *J. Org. Chem.*, **30**, 2635 (1965).

1-CYANOBENZOCYCLOBUTENE

(Bicyclo[4.2.0]octa-1,3,5-triene-7-carbonitrile)

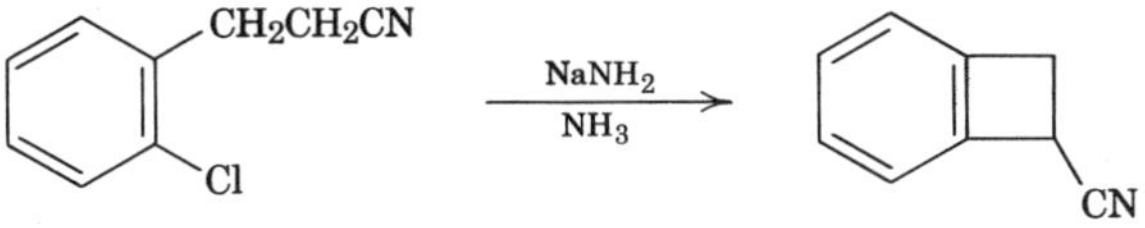

Submitted by J. A. Skorcz and F. E. Kaminski [1]
Checked by V. Z. Williams and K. B. Wiberg

1. Procedure

A 2-l. three-necked flask is thoroughly dried and fitted with a large dry-ice condenser, a mechanical stirrer, a nitrogen inlet, and a powder funnel in an efficient hood. With nitrogen flowing through the system, 62.5 g. (1.60 moles) of commercial sodium amide (Note 1) is added rapidly. (*Caution! Sodium amide is corrosive and readily decomposes in the presence of moisture.*) The funnel is replaced by a gas-inlet tube, the condenser is filled with a mixture of dry ice and acetone, and *ca.* 400 ml. of liquid ammonia is introduced from a cylinder. The gas-inlet tube is replaced by an addition funnel, stirring is commenced, and 66.3 g. (0.400 mole) of *o*-chlorohydrocinnamonitrile (Note 2) is added over a 10-minute period. The last traces of the nitrile are washed into the flask with small amounts of anhydrous ether.

The dark green reaction mixture is stirred vigorously for 3 hours and then is treated carefully with 96 g. (1.2 moles) of solid ammonium nitrate (Note 3). All the fittings are removed from the flask, and the ammonia is allowed to evaporate (Note 4). Water (300 ml.) is added cautiously to the residue. (*Caution! Traces of undecomposed sodium amide may adhere to the upper walls of the flask.*) The organic layer is taken up in two 160-ml. portions of chloroform, and the solutions are combined and washed twice with 100 ml. of 5% hydrochloric acid and once with 100 ml. of water. (*Caution! The extraction procedure and subsequent chloroform distillation should be conducted in a hood because some hydrogen cyanide is usually evolved.*) The chloroform solution is dried over anhydrous sodium sulfate, and the chloroform is removed by distillation. The residual liquid is distilled under reduced pressure through an insulated, 5-in. Vigreux column. The forerun, b.p. 95–100° (3 mm.), weighs *ca.* 1 g.; the product boils at 100–101° (3 mm.); n^{25}_{D} 1.5451. The yield of 1-cyanobenzocyclobutene is 33–34 g. (64–66%) (Notes 5 and 6).

2. Notes

1. The sodium amide was obtained from Farchan Research Laboratories and was approximately 90% pure.

2. The submitters prepared *o*-chlorohydrocinnamonitrile by the following procedure. Ethyl cyanoacetate (3040 g., 27 moles) was added to a solution of 140 g. (6.1 g. atoms) of sodium in 4 l. of absolute ethanol, followed by 970 g. (6 moles) of *o*,α-dichlorotoluene (Eastman Organic Chemicals), to afford 890 g. (63%) of ethyl 2-(*o*-chlorobenzyl)cyanoacetate,[2] b.p. 117–123° (0.03 mm.). Hydrolysis of this material in 2 l. of 10% aqueous sodium hydroxide at room temperature gave a quantitative yield (790 g.) of 2-(*o*-chlorobenzyl)cyanoacetic acid, m.p. 129–132° without recrystallization. Decarboxylation of 750 g. of the acid in 750 ml.

of refluxing dimethylformamide gave 550 g. (93%) of *o*-chlorohydrocinnamonitrile,[3] b.p. 82–85° (0.3 mm.), $n^{25}\text{D}$ 1.5362. The checkers carried out this preparation starting with 8 moles of ethyl cyanoacetate and obtained comparable yields.

3. Other ammonium salts, such as ammonium chloride, are equally satisfactory.

4. Overnight evaporation at room temperature is convenient.

5. The submitters carried out the reaction on 1-molar and 3-molar scales and obtained yields of 62–64% and 67%, respectively.

6. This procedure has also been used to obtain 1-cyano-5-methoxybenzocyclobutene from 2-bromo-4-methoxyhydrocinnamonitrile.[4]

3. Methods of Preparation

1-Cyanobenzocyclobutene has been prepared from sodium cyanide and 1-bromobenzocyclobutene,[5] formed by reaction of benzocyclobutene with N-bromosuccinimide,[6] and by ring closure of *o*-chlorohydrocinnamonitrile with potassium amide in liquid ammonia.[7] The present procedure is a modification of the latter method and was previously described by one of the submitters.[8]

4. Merits of the Preparation

Cyclization by addition of a side-chain carbanion to an aryne bond has been proposed as the method of choice for synthesis of the versatile 1-substituted benzocyclobutene system.[7] This general procedure now has been modified to permit convenient large-scale preparations utilizing a commercially available base, a minimum amount of liquid ammonia, and distillation for isolation of the product.

1. Lakeside Laboratories, Division of Colgate-Palmolive Company, Milwaukee, Wisconsin 53201.

2. P. E. Gagnon, J. L. Boivin, and J. Giguère, *Can. J. Res.*, **28B**, 352 (1950).
3. A. D. Grebenyuk and I. P. Tsukervanik, *Zh. Obshch. Khim.*, **25**, 286 (1955); *J. Gen. Chem. USSR (Engl. Transl.)*, **25**, 269 (1955).
4. J. A. Skorcz and J. E. Robertson, *J. Med. Chem.*, **8**, 255 (1965).
5. M. P. Cava, R. L. Litle, and D. R. Napier, *J. Am. Chem. Soc.*, **80**, 2257 (1958).

6. M. P. Cava and D. R. Napier, *J. Am. Chem. Soc.*, **80**, 2255 (1958).
7. J. F. Bunnett and J. A. Skorcz, *J. Org. Chem.*, **27**, 3836 (1962).
8. J. A. Skorcz, J. T. Suh, C. I. Judd, M. Finkelstein, and A. C. Conway, *J. Med. Chem.*, **9**, 656 (1966).

CYCLODECANONE

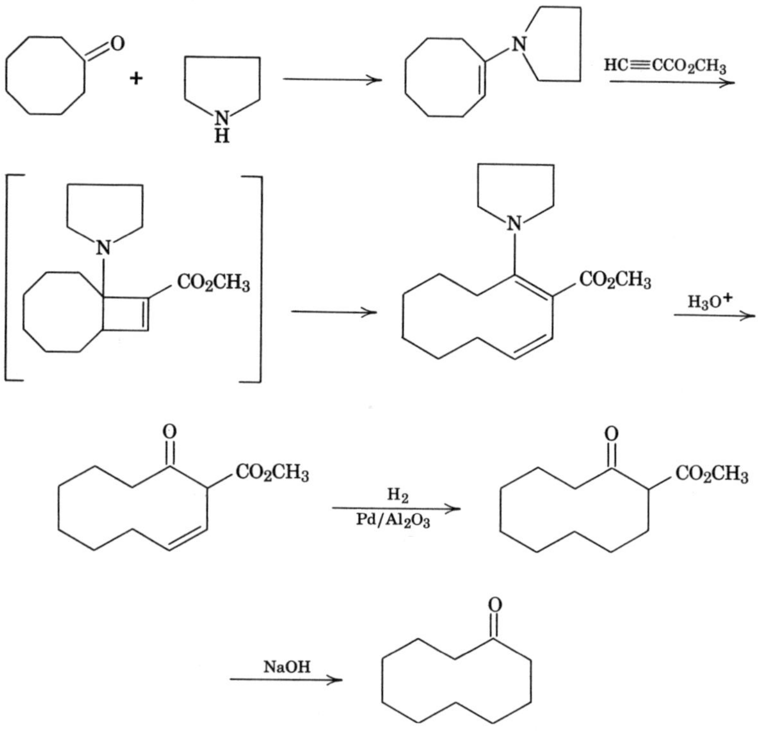

Submitted by R. D. Burpitt and J. G. Thweatt[1]
Checked by William G. Dauben, Michael H. McGann, and Noel Vietmeyer

1. Procedure

To a 500-ml. round-bottomed flask fitted with a 25- to 30-cm. column packed with glass helices to which is attached a water separator[2] filled with hexane (Note 1) are added 126 g. (1.00 mole) of cyclooctanone (Note 2), 100 g. (1.4 moles) of pyrrolidine, 100 ml. of xylene, and 0.5 g. of *p*-toluenesulfonic acid. The solution is heated under reflux until the separation of water ceases (Note 3). The water separator is replaced by a distillation head, and the reaction mixture is distilled through the column under reduced pressure to remove solvent and unreacted starting materials. When the head temperature reaches 50° (1 mm.), distillation is stopped, and the residue of almost pure N-(1-cyclo-octen-1-yl)pyrrolidine (152–161 g.) is used in the next step without further purification (Note 4).

The crude enamine is dissolved in 450 ml. of ether, and the solution is transferred to a 1-l. three-necked flask equipped with a sealed stirrer, a 250-ml. dropping funnel, and a two-necked adapter fitted with a calcium chloride tube and a thermometer immersed in the solution. A solution of 71–76 g. (0.85–0.90 mole) (Note 5) of methyl propiolate (*Caution! Methyl propiolate is a severe lachrymator and should be handled only in the hood.*) in 150 ml. of ether is added dropwise. During the addition the temperature of the mixture is maintained at 25–30° by periodic cooling of the reaction flask in a dry ice-acetone bath. When the addition is almost complete, a white solid begins to separate. The mixture is stirred at 25–30° for an additional hour, cooled to 0°, and filtered to remove the solid. This is dissolved in 700 ml. of 6% hydrochloric acid (Note 6), the acidic solution is warmed at 55–60° for 1 hour, and the mixture is cooled and extracted with two 100-ml. portions of ether. The ether is removed on a steam bath, and the residue of crude methyl 10-oxocyclodec-2-ene-1-carboxylate is dissolved in 300 ml. of methanol and hydrogenated over 5 g. of 5% palladium-on-alumina catalyst at 40 p.s.i. pressure and room temperature.

The catalyst is filtered, 200 g. (155 ml.) of 25% aqueous sodium hydroxide is added to the filtrate, and the mixture is heated under reflux for 1 hour. The condenser is replaced by a short Vigreux column and distillation head, and the heating is continued until most of the methanol has distilled. The two-phase residue is cooled and extracted with two 100-ml. portions of ether. The ether is removed on a steam bath, and the residue is distilled through a 20-cm. Vigreux column to yield 68–77 g. (44–50%) of cyclodecanone, b.p. 94–98° (10 mm.), m.p. 20–22° (Note 7).

2. Notes

1. If hexane is not used in the trap, an excessive amount of pyrrolidine is lost in the aqueous layer.

2. Cyclooctanone from Aldrich Chemical Co., methyl propiolate from Farchan Research Laboratories, and pyrrolidine from Eastman Organic Chemicals were used as received.

3. The reaction is usually complete after 3–6 hours at reflux. Owing to dissolved pyrrolidine, the aqueous layer amounts to 35–45 ml., and thus its volume is not a good measure of the extent of reaction.

4. Pure N-(1-cycloocten-1-yl)pyrrolidine, b.p. 76–78° (1 mm.), may be isolated by distillation through a Vigreux column.

5. The amount used should be adjusted to be equimolar with the amount of crude enamine.

6. This solid intermediate is reasonably stable to storage under nitrogen; however, the yield in the acid hydrolysis step is better when freshly prepared material is hydrolyzed immediately.

7. The same reaction sequence may be used to convert cyclododecanone to cyclotetradecanone. Preparation of the pyrrolidine enamine of cyclododecanone requires 2–3 days at reflux, and reaction of the enamine with methyl propiolate is best carried out in refluxing hexane. The enamine-propiolate reaction may also be used to convert cycloheptanone to cyclononanone. In this case the procedure must be modified to provide for partial hydrogenation of the intermediate amino ester without prior hydrolysis.[3] The reduced intermediate is saponified as described in the present procedure.

3. Methods of Preparation

Cyclodecanone has been obtained together with other products in the pyrolysis of the thorium or yttrium salts of nonanedioic acid.[4] It has also been prepared by reduction of sebacoin with zinc and hydrochloric acid,[5, 6] by dehydration of sebacoin followed by catalytic hydrogenation,[7] by ring enlargement of cyclononanone with diazomethane [8, 9] and of cyclooctanone with diazomethane in the presence of a Lewis acid catalyst,[9] by hydroboration of 1,2-cyclodecadiene followed by oxidation of the organoborane,[10] and by the present procedure.[3]

4. Merits of the Preparation

The chief merits of this preparation are its simplicity and the high purity of the product. Although the synthesis involves several steps, each step is a simple operation, and all intermediates may be used in the subsequent steps without purification. The purity of even the crude product is high, and any impurities which may be present are readily removed by a simple distillation.

The overall yield of cyclodecanone is comparable to the overall yield obtained by conversion of dimethyl sebacate to sebacoin [11] and subsequent reduction to cyclodecanone.[6] In addition, the present procedure does not require the use of a high-speed stirrer, the rigorous exclusion of air, and the high dilution that are necessary in preparing sebacoin.

1. Research Laboratories, Tennessee Eastman Company, Division of Eastman Kodak Company, Kingsport, Tennessee 37662.
2. S. Natelson and S. Gottfried, *Org. Syntheses*, Coll. Vol. **3**, 381 (1955).
3. K. C. Brannock, R. D. Burpitt, V. W. Goodlett, and J. G. Thweatt, *J. Org. Chem.*, **29**, 818 (1964).
4. L. Ruzicka, M. Stoll, and H. Schinz, *Helv. Chim. Acta*, **9**, 249 (1926); **11**, 670 (1928).
5. V. Prelog, L. Frenkiel, M. Kobelt, and P. Barman, *Helv. Chim. Acta*, **30**, 1741 (1947).
6. A. C. Cope, J. W. Barthel, and R. D. Smith, *Org. Syntheses*, Coll. Vol. **4**, 218 (1963).
7. M. Stoll, *Helv. Chim. Acta*, **30**, 1837 (1947).
8. E. P. Kohler, M. Tishler, H. Potter, and H. T. Thompson, *J. Am. Chem. Soc.*, **61**, 1057 (1939).
9. E. Müller, M. Bauer, and W. Rundel, *Tetrahedron Letters*, No. 13, 30 (1960).
10. D. Devaprabhakara and P. D. Gardner, *J. Am. Chem. Soc.*, **85**, 1458 (1963).
11. N. L. Allinger, *Org. Syntheses*, Coll. Vol. **4**, 840 (1963).

DIAMINOMALEONITRILE (HYDROGEN CYANIDE TETRAMER)

(Maleonitrile, diamino-)

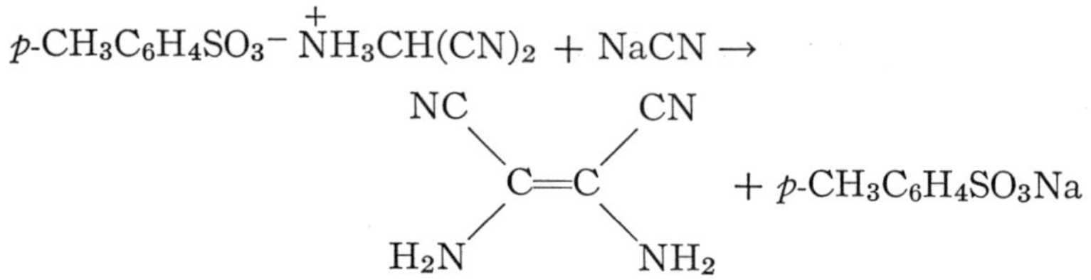

Submitted by J. P. Ferris and R. A. Sanchez[1]
Checked by O. W. Webster and R. E. Benson

1. Procedure

Caution! The preparation should be carried out in a hood because hydrogen cyanide may be evolved.

To a cooled (0°), stirred suspension of 10.0 g. (0.0395 mole) of aminomalononitrile *p*-toluenesulfonate[2] in 20 ml. of water is added 10.0 g. (0.204 mole) of sodium cyanide in 30 ml. of ice water. One minute (Note 1) after the addition of the sodium cyanide the precipitated product is collected by filtration and washed with 20 ml. of ice water. The solid is immediately dissolved (Note 1) in 30 ml. of boiling isobutyl alcohol, and the solution is stirred with 0.4 g. of Darco activated carbon (Note 2). The mixture is filtered rapidly through 10 g. of Celite filter aid, and the filter cake is washed with 10 ml. of hot isobutyl alcohol. The product that crystallizes on cooling is collected by filtration and washed with 10 ml. of isobutyl alcohol to give 0.95–1.1 g. (22–26%) of white needles, m.p. 181–183° (dec.).

2. Notes

1. The product darkens on long standing.
2. Hydrogen cyanide tetramer is strongly adsorbed on acti-

vated carbon; no more than the recommended amount of carbon should be used, and it should be added carefully to avoid frothing.

3. Methods of Preparation

The present procedure is a modification of the original synthesis.[3] Hydrogen cyanide tetramer can be prepared directly from hydrogen cyanide.[4]

4. Merits of the Preparation

This is a convenient laboratory preparation of hydrogen cyanide tetramer that avoids the hazards in using hydrogen cyanide itself. Hydrogen cyanide tetramer is a useful intermediate for the synthesis of heterocycles such as imidazoles[5, 6] and thiadiazoles.[7]

1. The Salk Institute for Biological Studies, San Diego, California [Present address (J.P.F.): Department of Chemistry, Rensselaer Polytechnic Institute, Troy, New York 12181].

2. J. P. Ferris, R. A. Sanchez, and R. W. Mancuso, *Org. Syntheses*, this volume, p. 1.
3. J. P. Ferris and L. E. Orgel, *J. Am. Chem. Soc.*, **88**, 3829 (1966); **87**, 4976 (1965).
4. H. Bredereck, G. Schmötzer, and E. Oehler, *Ann.*, **600**, 81 (1956).
5. H. Bredereck and G. Schmötzer, *Ann.*, **600**, 95 (1956).
6. J. P. Ferris and L. E. Orgel, *J. Am. Chem. Soc.*, **88**, 1074 (1966).
7. M. Carmack, L. M. Weinstock, and D. Shew, Abstracts, 136th National Meeting of the American Chemical Society, Atlantic City, N. J., Sept. 1959, p. 37P; D. Shew, *Dissertation Abstr.*, **20**, 1593 (1959).

DI-*t*-BUTYL NITROXIDE

(Nitroxide, di-*tert*-butyl)

$$(CH_3)_3CNO_2 \xrightarrow[2.\ H_2O]{1.\ Na} [(CH_3)_3C]_2NO$$

Submitted by A. K. HOFFMANN, A. M. FELDMAN, E. GELBLUM, and A. HENDERSON [1]
Checked by R. A. HAGGARD and WILLIAM D. EMMONS

1. Procedure (Note 1)

Thirteen hundred milliliters of 1,2-dimethoxyethane (glyme) is distilled from lithium aluminum hydride (*Caution! Under no circumstances should distillation be carried to dryness since explosive decomposition of the residual hydride may occur.*) (Note 2) directly into a 2-l., nitrogen-flushed, three-necked, Morton flask equipped with a nitrogen inlet, an outlet, and a high-speed stirrer with a stainless steel propeller-type blade. The flask is charged with 89.7 g. (0.87 mole) of *t*-nitrobutane (Note 3) and 19.9 g. (0.87 g. atom) of sodium cut into pea-sized pieces (Note 4). The stirrer is started and initially operated at a speed just adequate to draw some of the sodium through the blade. The onset of reaction is signaled when the solution is pale lavender and the sodium surface is clearly etched and colored bright gold (Note 5). The temperature of the reaction mixture is maintained at 25–30° (Note 6) by directing an air blast at the sides of the flask and by controlling the rate at which sodium is drawn through the blades of the stirrer. As the reaction progresses, colorless solid is formed, and at the end of the reaction (*ca.* 24 hours) the reaction mixture consists of solid and a colorless glyme solution. Most of the glyme is removed by evaporation under reduced pressure at room temperature with a water bath at 20–25° to leave a thick, colorless slurry. To the slurry under nitrogen (Note 7) is added 270 ml. of water, and the reddish brown organic layer is separated. The aqueous layer (Note 8) is extracted with several portions of pentane until the extract is colorless. The organic layer and pentane extracts are combined, cooled to 0°, and washed rapidly and thoroughly with two 70-ml. portions of ice-cold 0.25*N* hydrochloric acid to remove hydroxylamine impurities (Note 9). The pentane solution of the product is washed immediately with 70 ml. of cold water followed by 70 ml. of cold, aqueous 0.2*N* sodium hydroxide. The combined, cold, aqueous acidic washings are extracted with small portions of pentane until colorless. This pentane extract is used to extract the aqueous sodium hydroxide layer and is then washed with water and combined with the initial pentane extract.

The pentane solution is dried over anhydrous magnesium sulfate and fractionated with an efficient spinning-band column (Note 10). After foreruns of pentane and glyme containing *t*-nitrosobutane are removed (Note 11), 26–27 g. (42–43%) of red di-*t*-butyl nitroxide, b.p. 59–60° (11 mm.), is obtained.

2. Notes

1. The submitters obtained similar results using a preparative scale 7.5 times that described here; yield 36%.

2. The 1,2-dimethoxyethane (Ansul Chemical Co.) was pre-dried for several days over calcium hydride, filtered, and stored over lithium aluminum hydride prior to its distillation at atmospheric pressure immediately before use. For a larger-scale preparation it is expeditious to distil simultaneously from two 5-l. flasks rather than from a single large one. Under these conditions, distillation of the glyme can be completed in 8–10 hours.

3. The *t*-nitrobutane employed was prepared by the procedure of Kornblum, Clutter, and Jones.[2] This method is essentially the same as that previously reported in *Organic Syntheses*.[3]

4. Throughout all transfers, air must be rigorously excluded from the flask by the nitrogen blanket.

5. When great care has not been taken to ensure the absence of moisture, induction times as long as several minutes are observed before the onset of reaction.

6. The temperature of the reaction mixture must never be

allowed to exceed 30°; above this temperature drastic diminution of yield occurs.

7. A nitrogen blanket is used here to prevent ignition of hydrogen resulting from traces of unreacted sodium.

8. At this point during one run the checkers obtained a considerable amount of water-insoluble, colorless solid; however, the product yield was not changed.

9. It is essential to keep the reaction mixture and acid cold since otherwise substantial decomposition of product results.

10. The pot temperature should not exceed 100° until the di-*t*-butyl nitroxide fraction is collected.

11. In several runs it was noted that, despite the acid extraction, small amounts of N,N-di-*t*-butylhydroxylamine crystallized in the cooler parts of the fractionating column head. In such cases, repetition of the acid extraction procedure is required before fractionation.

3. Methods of Preparation

The procedure described is that of Hoffmann, Feldman, Gelblum, and Hodgson [4] and is the only one known at this time for the preparation of substantial amounts of di-*t*-butyl nitroxide.

4. Merits of the Preparation

The method is specific for the preparation of di-*t*-butyl nitroxide, a liquid member of a group of stable free radicals useful for the inhibition of a variety of reactions proceeding by radical chain mechanisms as well as for providing standards for e.s.r. measurements.

1. American Cyanamid Company, Stamford Research Laboratories, 1937 West Main Street, Stamford, Connecticut 06904.
2. N. Kornblum, R. J. Clutter, and W. J. Jones, *J. Am. Chem. Soc.*, **78**, 4003 (1956).
3. N. Kornblum and W. J. Jones, *Org. Syntheses*, **43**, 87 (1963).
4. A. K. Hoffmann, A. M. Feldman, E. Gelblum, and W. G. Hodgson, *J. Am. Chem. Soc.*, **86**, 639 (1964).

1,3-DIHYDROISOINDOLE

(Isoindoline)

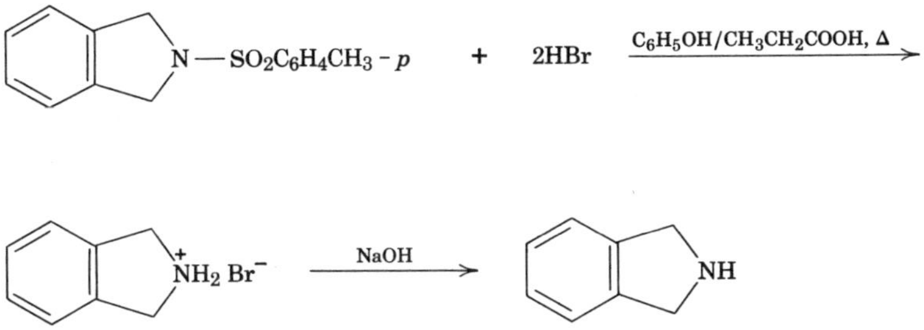

Submitted by J. BORNSTEIN, J. E. SHIELDS,
and A. P. BOISSELLE [1]
Checked by WILLIAM G. DAUBEN and HAROLD B. MORRIS

1. Procedure

In a 1-l. round-bottomed flask are placed 36.0 g. (0.132 mole) of 2-(*p*-tolylsulfonyl)dihydroisoindole,[2] 36.0 g. (0.38 mole) of phenol, 270 ml. of 48% hydrobromic acid (Note 1), and 45 ml. of propionic acid. A few boiling chips are added, and the flask is fitted with a reflux condenser in the top of which is placed a T-tube connected to a source of low-pressure nitrogen and to a mercury bubbler. The mixture is heated under reflux for 2 hours in an atmosphere of nitrogen. The deeply colored reaction mixture is cooled to room temperature, transferred to a 1-l. separatory funnel, and washed with two 200-ml. portions of ether (Note 2). The aqueous phase is then added dropwise over a 1-hour period to a vigorously stirred (Note 3) solution of 200 g. of sodium hydroxide in 600 ml. of water in a 2-l. Erlenmeyer flask immersed in an ice bath. The solution is transferred to a 3-l. separatory funnel and extracted with five 300-ml. portions of ether. The ethereal extracts are combined, dried over anhydrous potassium carbonate (Note 4), and filtered. The solvent is distilled, and the dark residual oil is transferred to a distillation flask and distilled through a low-holdup, semimicro column (Note 5). After removal of 1 or 2 drops of forerun, colorless 1,3-dihydroisoindole is collected at 96–97° (10 mm.) or 55–56° (2 mm.), n^{25}D 1.5686, d_4^{20} 1.081. The yield is 9.9–11.2 g. (63–71%) (Note 6).

2. Notes

1. The hydrobromic acid should be colorless. Reagent grade 47–49% hydrobromic acid, obtained from J. T. Baker Chemical Co., was used as supplied. A technical grade of the constant-boiling acid is suitable if purified by distillation from stannous chloride.

2. Detection of the water-ether interface may prove troublesome; backlighting of the separatory funnel by an intense light source is recommended. The same volume of ether must be used for each washing, even when the preparation is carried out on a smaller scale, *e.g.*, one-half or one-third the scale described here.

3. Stirring is most conveniently accomplished with a magnetic stirrer.

4. Washing of the ethereal extract with water decreases the yield of product.

5. The submitters used a 7 × 300-mm. externally heated column packed with a helix of Chromel wire and fitted with a partial reflux head.[3]

6. Since the product slowly darkens on exposure to air, it should be stored under nitrogen in a refrigerator. The compound solidifies on cooling; m.p. 16.0–16.5°. Nuclear magnetic resonance spectrum (neat, tetramethylsilane internal standard): singlets at δ 7.00 (aromatic protons), 3.93 (CH_2), and 2.24 p.p.m. (NH).

3. Methods of Preparation

1,3-Dihydroisoindole has been prepared from phthalimide by electrolytic reduction[4] and by reduction with lithium aluminum hydride.[5] Other methods that have been used are reduction of 1-chlorophthalazine with zinc and hydrochloric acid[6] and hydrogenolysis of 2-benzyl-1,3-dihydroisoindole.[7] The present method is essentially that of Bornstein, Lashua, and Boisselle.[8]

4. Merits of the Preparation

This procedure illustrates a general method for the preparation of amines by reductive cleavage of sulfonamides by hydrobromic acid in the presence of phenol.[9] The present synthesis makes 1,3-dihydroisoindole readily accessible and is superior in certain respects to the other two practical methods of preparation. Thus the method here described is shorter and gives a higher overall yield than the three-step synthesis of Neumeyer,[7] and obviates the special apparatus and careful control required by the electrochemical process of Dunet, Rollet, and Willemart.[4]

1. Department of Chemistry, Boston College, Chestnut Hill, Massachusetts 02167.
2. J. Bornstein and J. E. Shields, *Org. Syntheses*, **47**, 110 (1967).
3. C. W. Gould, Jr., G. Holzman, and C. Niemann, *Anal. Chem.*, **20**, 361 (1948).
4. A. Dunet, J. Rollet, and A. Willemart, *Bull. Soc. Chim. France*, [5] **17**, 877 (1950).
5. A. Uffer and E. Schlittler, *Helv. Chim. Acta*, **31**, 1397 (1948).
6. S. Gabriel and A. Neumann, *Ber.*, **26**, 521 (1893).
7. J. L. Neumeyer, *J. Pharm. Sci.*, **53**, 981 (1964).
8. J. Bornstein, S. C. Lashua, and A. P. Boisselle, *J. Org. Chem.*, **22**, 1255 (1957).
9. S. Searles and S. Nukina, *Chem. Rev.*, **59**, 1077 (1959).

7,7-DIMETHOXYBICYCLO[2.2.1]HEPTENE

(2-Norbornen-7-one dimethyl acetal)

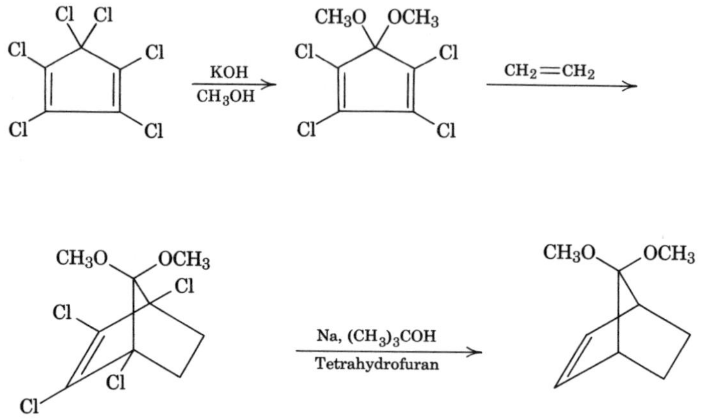

Submitted by P. G. Gassman and J. L. Marshall [1]
Checked by William G. Dauben and James L. Chitwood

1. Procedure

Caution! Most polychlorinated compounds show some toxicity. These compounds should be handled in a hood.

A. *5,5-Dimethoxy-1,2,3,4-tetrachlorocyclopentadiene.* In a 3-l. three-necked flask fitted with a condenser (Note 1), an addition funnel, and a mechanical stirrer (Note 2) are placed 254 g. (0.93 mole) of hexachlorocyclopentadiene (Note 3) and 800 ml. of methanol (Note 4). The stirrer is started, and a solution of 120 g. (2.14 moles) of potassium hydroxide in 600 ml. of methanol is added dropwise over a period of 2 hours (Note 5). The reaction mixture is stirred for an additional 2 hours and then poured over 3 l. of chopped ice. After the ice has melted, the mixture is extracted with three 250-ml. portions of dichloromethane. The combined extracts are dried over anhydrous magnesium sulfate and concentrated to a yellow syrup on a rotary evaporator

(Note 6). The residue is distilled through a 12-in. vacuum-jacketed Vigreux column to yield 187–189 g. (76–77%) of 5,5-dimethoxy-1,2,3,4-tetrachlorocyclopentadiene as a viscous, yellow-tinted oil, b.p. 79–84° (0.6 mm.) (Note 7).

B. *7,7-Dimethoxy-1,2,3,4-tetrachlorobicyclo*[*2.2.1*]*hept-2-ene.* A large Pyrex gas washing bottle with fritted-glass inlet (Note 8) is fitted with a condenser and a drying tube. In the bottle is placed 189 g. (0.72 mole) of 5,5-dimethoxy-1,2,3,4-tetrachlorocyclopentadiene, and a slow stream of nitrogen and ethylene is passed through the fritted-glass inlet (Note 9). The bottle is heated to 180–190° by means of an oil bath. The color of the liquid changes from yellow to reddish brown as ethylene is bubbled through the reaction mixture at this temperature for 6 hours (Note 10). The reaction mixture is cooled and distilled through a 12-in. vacuum-jacketed Vigreux column to yield 155–165 g. (73–78%) of a yellow syrup, b.p. 70–75° (0.15 mm.) (Note 11).

C. *7,7-Dimethoxybicyclo*[*2.2.1*]*heptene.* A 3-l. three-necked flask is equipped with a sealed Hershberg stirrer,[2] a condenser fitted with a nitrogen inlet to maintain a slight positive pressure, and a pressure-equalizing dropping funnel. The flask is placed in a heating mantle, and into it are placed 1.5 l. of tetrahydrofuran, 130 g. (5.7 g. atoms) of sodium chopped into 5-mm. cubes, and 190 ml. (150 g., 2.0 moles) of *t*-butyl alcohol. This mixture is stirred vigorously and brought to gentle reflux (Note 12). As soon as refluxing occurs, 106 g. (0.36 mole) of 7,7-dimethoxy-1,2,3,4-tetrachlorobicyclo[2.2.1]hept-2-ene is added dropwise over a 2-hour period (Note 13). The mixture is heated under reflux for 38 hours, cooled to room temperature, and filtered through a wire screen to remove the unreacted sodium. The dark filtrate is refiltered by suction through Celite in a Buchner funnel (Note 14). The filtrate is mixed with 2 l. of chopped ice and 500 ml. of ether. The aqueous phase is separated (Note 15), and the organic phase is washed with 500-ml. portions of saturated aqueous sodium chloride until the washings are clear. The ethereal solution is dried over anhydrous magnesium sulfate and concentrated to a dark oil by removal of the ether by fractional distillation. The oil is fractionally distilled through a 6-in. Vigreux column to yield 17–24 g. (31–43%) of colorless liquid, b.p. 58–68° (17 mm.), n^{25}_{D} 1.4598 (Notes 16 and 17).

2. Notes

1. If the directions are carefully followed, the condenser will not be utilized since it serves mainly as a safety device in case the reaction should become too exothermic.

2. A Hershberg nichrome wire stirrer[2] is well suited for this reaction.

3. The hexachlorocyclopentadiene was used as obtained from Matheson, Coleman and Bell.

4. Commercial grade methanol was used.

5. This reaction mixture should not be cooled initially because an uncontrollable exothermic reaction will occur if a large concentration of alkoxide builds up.

6. Concentration at 100° (30 mm.) is necessary for removal of most of the dichloromethane.

7. The submitters have obtained an 86% yield of product, b.p. 79–91° (0.6 mm.). They have also found that the reaction may be scaled up fivefold if 4 hours is taken for the addition of the methanolic base. No danger exists if the temperature is maintained between 50° and 60°.

8. Pyrex gas washing bottle, Corning No. 31750, was used.

9. The checkers found that best results were obtained when the slowest detectable nitrogen flow was used with a fairly rapid ethylene flow (about 1 in. of foam in the gas washing bottle at the reaction temperature).

10. The course of the reaction is readily followed by n.m.r. spectroscopy. The spectrum of the starting material has a singlet at δ 3.30 p.p.m., whereas that of the product has two singlets at δ 3.50 and 3.55 p.p.m. The time required for complete reaction depends on the flow rate of ethylene and nitrogen. The reaction should be allowed to continue until all the starting material is consumed.

11. The distillation should be conducted carefully since the

yield in the next step depends on the purity of the material used.

12. Occasionally the pieces of sodium may start to fuse together. This difficulty may be avoided by bringing the mixture to a gentle reflux and stirring vigorously. Once the addition of the chlorinated compound is started, fusing of the sodium ceases. When the reaction is finished, the sodium pieces often fuse into a single large chunk.

13. Unless the reaction mixture is heated *before* the addition process is started, there may be an initial induction period that may cause the reaction to become extremely vigorous once the mixture heats to reflux temperature.

14. The checkers found this last filtration to be a very time-consuming and cumbersome operation. They found it preferable to omit it; they cautiously added methanol [3] to decompose any traces of sodium and poured the resulting solution directly onto the chopped ice.

15. In the separation of the organic and aqueous phases it is often initially very difficult to discern the phase separation because of the dark color of the reaction mixture. The submitters found that, if the phase separation cannot be detected under ordinary light, it can usually be seen by the use of an ultraviolet scanning lamp.

16. The submitters report that 65% yields of product, b.p. 61–71° (18 mm.), can be obtained by workers with experience with this reaction.

17. The product is rather volatile, and care should be taken in its handling and storing.

3. Methods of Preparation

The procedures described for the preparation of 5,5-dimethoxy-1,2,3,4-tetrachlorocyclopentadiene and 7,7-dimethoxy-1,2,3,4-tetrachlorobicyclo[2.2.1]hept-2-ene are essentially those of Newcomer and McBee [4] and of Hoch,[5] respectively. The dechlorination is a modification of an analogous dechlorination carried out by Bruck, Thompson, and Winstein.[3] The overall procedure is that of Gassman and Pape.[6]

4. Merits of the Preparation

The reactions include an unusual Diels-Alder reaction and a very useful synthetic method, the dechlorination of polychlorinated compounds. At the present time this procedure is the best one available for the removal of chlorine from an organic molecule. The end product, 7,7-dimethoxybicyclo[2.2.1]heptene, is an interesting and useful intermediate in bicyclic chemistry; it has a reactive double bond and a protected carbonyl group in the 7-position.

1. Department of Chemistry, The Ohio State University, Columbus, Ohio 43210.
2. P. S. Pinkney, *Org. Syntheses*, Coll. Vol. **2**, 116 (1943).
3. P. Bruck, D. Thompson, and S. Winstein, *Chem. Ind.* (*London*), 405 (1960).
4. J. S. Newcomer and E. T. McBee, *J. Am. Chem. Soc.*, **71**, 946 (1949).
5. P. E. Hoch, *J. Org. Chem.*, **26**, 2066 (1961).
6. P. G. Gassman and P. G. Pape, *J. Org. Chem.*, **29**, 160 (1964).

DIMETHYLKETENE β-LACTONE DIMER

(3-Pentenoic acid, 3-hydroxy-2,2,4-trimethyl-, β-lactone)

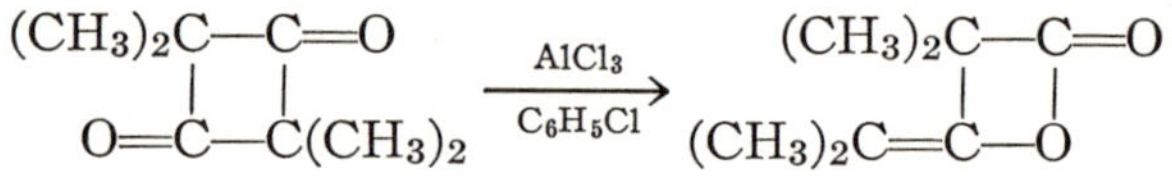

Submitted by ROBERT H. HASEK,[1] R. DONALD CLARK,[1] and GERALD L. MAYBERRY[2]
Checked by V. BOEKELHEIDE, J. WITTE, and G. SINGER

1. Procedure

Caution! Dimethylketene β-lactone dimer is a mild but deceptively persistent lachrymator.

In a 500-ml. three-necked flask equipped with a thermometer, a mechanical stirrer, and a reflux condenser are placed 200 g. (1.43 moles) of tetramethyl-1,3-cyclobutanedione (Note 1) and 50 g. of chlorobenzene (Note 2). The mixture is heated to 135° with stirring while a total of 1.8 g. of reagent grade anhydrous aluminum chloride is added in 0.3-g. portions over a 3-hour period (Note 3). After the addition is complete, heating is continued for an additional 5 hours (Note 4).

The reaction mixture is then cooled to 35–40° and poured into a stirred solution of 230 g. of sodium chloride and 6.0 g. of sodium acetate in 600 ml. of water at 40°. The mixture is stirred for 15 minutes and then is transferred to a separatory funnel; the layers are separated, and the lower, aqueous layer is discarded. The crude product is distilled at reduced pressure through a stainless steel spinning-band column (Note 5). The yield of the β-lactone dimer of dimethylketene is 122–132 g. (61–67%); b.p. 69–71.5° (14 mm.) (Note 6). The product may be redistilled and the fraction boiling at 119.5–120° (150 mm.), $n^{20}D$ 1.4380, collected.

2. Notes

1. Tetramethyl-1,3-cyclobutanedione is available from Eastman Organic Chemicals. It melts at 115–116° and is typically 99% pure by vapor-phase chromatography.

2. More highly chlorinated aromatic solvents such as 1,2,4-trichlorobenzene can be used with similar results.

3. No special drying precautions are required; however, traces of water can be conveniently removed before the catalyst is added by distilling a small portion of the solvent until cloudiness disappears. The checkers found that the distillation of part of the solvent as indicated was necessary in order to obtain satisfactory yields. When the catalyst is added incrementally, no appreciable exotherm is observed.

4. The isomerization is usually complete in 5 hours and can easily be followed by vapor-phase chromatography. Heating periods up to 20 hours are not detrimental. The only failure among numerous preparations occurred when tetramethyl-1,3-cyclobutanedione contaminated with 4% of isobutyric acid was used. In case of partial conversion after 5 hours, additional increments (0.5 g.) of aluminum chloride should be added to complete the reaction.

5. The product may be distilled directly from the crude reaction mixture after addition of sodium acetate. The results are similar.

6. The distilled product is 99% pure by vapor-phase chromatography.

3. Methods of Preparation

The β-lactone dimer of dimethylketene can be prepared by pyrolysis of its polyester, which is formed by the base-catalyzed polymerization of dimethylketene.[3-5] In addition to the rearrangement of the normal dimer described above,[6] the direct dimerization of dimethylketene in the presence of aluminum chloride [3] or trialkyl phosphites [7] leads to the β-lactone dimer.

4. Merits of the Preparation

The β-lactone dimer of dimethylketene reacts with alcohols, phenols, mercaptans, and amines to form derivatives of 2,2,4-trimethylvaleric acid.[3] In this respect it is a more powerful acylating reagent than the normal dimer, tetramethyl-1,3-cyclobutanedione. The preparation of 2,2,4-trimethyl-3-oxovaleranilide, for example, is accomplished easily with the lactone dimer, but is extremely difficult with the normal dimer.[8]

In the presence of catalytic amounts of sodium methoxide, dimethylketene β-lactone dimer is polymerized at moderate temperature to a polyester.[3] At higher temperatures (above 100°), disproportionation to the cyclic trimer, hexamethyl-1.3,5-cyclohexanetrione, takes place.[9] Addition of a stoichiometric amount of sodium methoxide to the lactone dimer generates the sodium enolate of methyl 2,2,4-trimethyl-3-oxovalerate. This reaction provides a convenient entry into certain ester anion chemistry that formerly required the use of a strong base like tritylsodium.[10]

Although these reactions can be duplicated in most cases with the normal dimer of dimethylketene,[11] the more reactive lactone dimer is the preferred reagent. The liquid form of this dimer is convenient to handle. A distinct difference in behavior of the dimethylketene dimers is noted when they are pyrolyzed. The normal dimer is dissociated at 600° to dimethylketene,[12] but the lactone dimer is decarboxylated almost quantitatively at 450° to tetramethylallene.[13]

1. Research Laboratories, Tennessee Eastman Co., Kingsport, Tennessee 37662.
2. Organic Chemicals Division, Tennessee Eastman Co., Kingsport, Tennessee 37662.
3. R. H. Hasek, R. D. Clark, E. U. Elam, and J. C. Martin, *J. Org. Chem.*, **27**, 60 (1962).
4. G. Natta, G. Mazzanti, G. Pregaglia, M. Binaghi, and M. Peraldo, *J. Am. Chem. Soc.*, **82**, 4742 (1960).
5. G. Natta, G. Mazzanti, G. Pregaglia, and M. Binaghi, *Makromol. Chem.*, **44–46**, 537 (1961).
6. R. D. Clark, U. S. Patent 3,062,837 (1962) [*C.A.*, **58**, 8911 (1963)].
7. E. U. Elam, *J. Org. Chem.*, **32**, 215 (1967).
8. H. Staudinger and J. Mayer, *Ber.*, **44**, 528 (1911).
9. R. H. Hasek, R. D. Clark, E. U. Elam, and R. G. Nations, *J. Org. Chem.*, **27**, 3106 (1962).
10. Cf. C. R. Hauser and B. E. Hudson, Jr., *Org. Reactions*, **1**, 266 (1942).
11. R. H. Hasek, E. U. Elam, J. C. Martin, and R. G. Nations, *J. Org. Chem.*, **26**, 700 (1961).
12. W. E. Hanford and J. C. Sauer, *Org. Reactions*, **3**, 136 (1946).
13. J. C. Martin, U. S. Patent 3,131,234 (1964) [*C. A.*, **61**, 2969 (1964)].

1,1-DIPHENYLCYCLOPROPANE

(Cyclopropane, 1,1-diphenyl-)

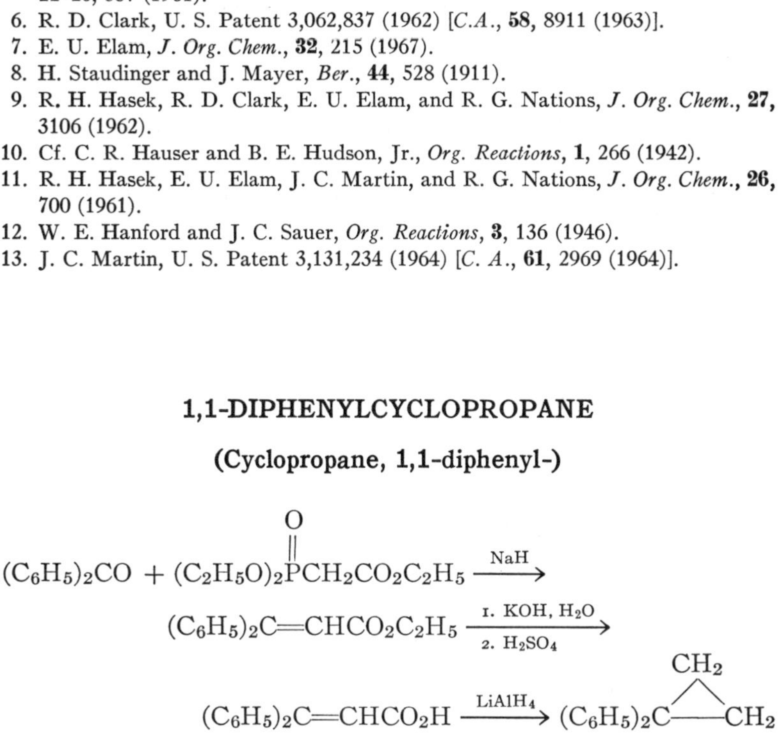

Submitted by M. J. JORGENSON and A. F. THACHER[1]
Checked by JOHN J. MILLER and WILLIAM D. EMMONS

1. Procedure

A. *β-Phenylcinnamic acid.* A suspension of 8.8 g. (0.20 mole) of a 55.1% sodium hydride dispersion in mineral oil (Note 1) and 300 ml. of 1,2-dimethoxyethane (Note 2) are added to a dry, 1-l., three-necked flask equipped with a stirrer, a condenser with drying tube, and a pressure-equalizing dropping funnel. The flask is immersed in an ice bath, and 44.8 g. (0.20 mole) of triethyl phosphonoacetate (Note 3) is added through the dropping funnel over a 20-minute period. After the addition the solution is stirred at room temperature for 30 minutes, and then 36.4 g. (0.20 mole) of benzophenone is added in one portion. The solution is heated at reflux for 4 days. The two-phase reaction mixture is cooled, and the flask is filled with water. The ester is extracted from the solution with three 100-ml. portions of ether. The ethereal extracts are combined, dried over anhydrous sodium sulfate, and evaporated to as small a volume as possible on the steam bath. The resulting mixture of ester and unreacted benzophenone is added to 140 ml. of water, 45 ml. of dioxane, and 26.4 g. (0.40 mole) of 85% potassium hydroxide in a 500-ml. one-necked flask, and the mixture is refluxed overnight (17 hours). The solution is cooled and extracted with 50 ml. of ether to remove benzophenone and any unhydrolyzed ester. The aqueous fraction is acidified with 10*N* sulfuric acid, and the resulting solid product is isolated by filtration, washed with two 50-ml. portions of water, and dried. Recrystallization from 200 ml. of benzene gives 28.3–33.9 g. (63–75%) of β-phenylcinnamic acid, m.p. 161–163° (Notes 4 and 5).

B. *1,1-Diphenylcyclopropane.* A 500-ml. three-necked flask is equipped with a nitrogen-inlet tube, a condenser, and a dropping funnel. The flask is charged with 100 ml. of dry tetrahydrofuran (Note 6) and 5.13 g. (0.135 mole) of lithium aluminum hydride (Note 1), and the system is purged with dry nitrogen. To the stirred solution is added dropwise over a period of 15–20 minutes a solution of 20 g. (0.089 mole) of β-phenylcinnamic acid (Note 7) in 100 ml. of tetrahydrofuran. The addition results in an exothermic reaction which causes the solution to reflux. The refluxing is allowed to subside after addition is complete, and then the mixture is carefully reheated to reflux. A mild exotherm occurs a few minutes after reflux is reached. This is easily controlled by removing the heat source or by cooling with ice. After the occurrence of this exothermic reaction, which is accompanied by a color change from rust to deep red, the mixture is heated at the reflux temperature for an additional 2 hours. The mixture is then cooled in an ice bath, and sufficient 10% sulfuric acid (*ca.* 5 ml.) is carefully (*Caution! Vigorous evolution of hydrogen.*) added dropwise to destroy excess lithium aluminum hydride. Another 100 ml. of 10% sulfuric acid is added after decomposition of the hydride is complete to dissolve the aluminum salts (Note 8). The acidified solution is worked up immediately (Note 9) by dilution with 200 ml. of water and extraction with one 100-ml. portion and three 50-ml. portions of ether. The combined ethereal extracts are washed with three 50-ml. portions of saturated aqueous sodium bicarbonate. The bicarbonate extracts are combined and washed with 50 ml. of ether, and all the ethereal extracts are combined. The ethereal solution is dried over anhydrous sodium sulfate, and the ether is removed at atmospheric pressure. The residue is distilled under reduced pressure to give 10.0–10.7 g. (57–62%) of diphenylcyclopropane (Note 10), b.p. 132–134° (10 mm.), $n^{20}\text{D}$ 1.590.

2. Notes

1. Sodium hydride dispersions in mineral oil and lithium aluminum hydride are available from Metal Hydrides, Inc.

2. Before use, 1,2-dimethoxyethane (ethylene glycol dimethyl ether) was partially dried over anhydrous calcium chloride and then distilled from lithium aluminum hydride. It was stored over sodium ribbon.

3. Triethyl phosphonoacetate is available from Aldrich Chemical Co.

4. About half of the solid is insoluble in the hot benzene and is removed by filtration.

5. The submitters obtained 28.2 g. (63%) of β-phenylcinnamic acid, m.p. 162–163°.

6. Tetrahydrofuran is best dried according to the procedure described in *Org. Syntheses*, **46**, 105 (1966). (*Caution! Note warning of danger in drying tetrahydrofuran containing peroxides.*) The checkers employed a peroxide-free grade of anhydrous tetrahydrofuran which is available from Fisher Scientific Co.

7. In the preparation of other phenylcyclopropanes by this method, esters rather than acids were employed. Use of the ester in the present preparation gave a poor yield of product.

Since the acid can easily be purified, its use here is also more expedient.

8. The checkers found this amount of acid insufficient to dissolve all of the salts, and an additional 5 ml. of concentrated sulfuric acid was added.

9. In general it has been found that yields of cyclopropanes are lowered if the acidic solutions are permitted to stand before workup.

10. Fractionation is not essential because the main contaminant is high-boiling polymer. The checkers obtained two fractions: b.p. 138–142° (11 mm.), 10.0–10.2 g., 99% pure by vapor-phase chromatography; and b.p. 142–148° (11 mm.), 0.6–0.9 g., 96% pure.

3. Methods of Preparation

1,1-Diphenylcyclopropane has been prepared in 24% yield by the Simmons-Smith reaction,[2] in 78% yield by treatment of 3,3-diphenylpropyltrimethylammonium iodide with sodium or potassium amide,[3] in 61% yield by reaction of 1,1-diphenylethylene with dimethylsulfonium methylide,[4] and in unspecified yields from 1,1-diphenylethylene by reaction with diazomethane followed by pyrolysis of the resulting pyrazoline or by reaction with ethyl diazoacetate followed by distillation of the corresponding acid over calcium oxide.[5]

β-Phenylcinnamic acid has been prepared previously by a variety of methods, the best of which appear to be the dehydration of ethyl β-hydroxy-β,β-diphenylpropionate by treatment with sodium acetate in acetic acid[6] and the reaction of 1,1-diphenylethylene with oxalyl chloride.[7]

4. Merits of the Preparation

This procedure illustrates a general method for the preparation of phenylcyclopropanes from cinnamic acids, esters, aldehydes, or alcohols.[8] It complements the Simmons-Smith reaction as a general method for the preparation of such cyclopropanes. It offers advantages over the Simmons-Smith method in cases in which electron-withdrawing substituents in the benzene ring or steric crowding around the double bond lead to low yields in the Simmons-Smith reaction. Also, in the case of possible stereoisomerism in the starting material or product, the present method leads stereospecifically to a single cyclopropane. It has the disadvantage that reducible substituents on the benzene ring do not survive the reductive treatment. The present method is an exceptionally simple, one-step preparative process employing starting materials that are commercially available, or readily accessible from aldehydes and ketones via the phosphono ester addition procedure of Wadsworth and Emmons.[9] Cyclopropane formation is particularly facile when electron-withdrawing substituents are present on the aromatic ring or when the β-position is substituted by another aryl group. Alkyl substitution on the double bond also facilitates cyclopropane formation. The examples recorded in Table I illustrate these effects. An excess of hydride is necessary for producing good yields of cyclopropanes.

TABLE I

PREPARATION OF PHENYLCYCLOPROPANES FROM CINNAMIC ESTERS

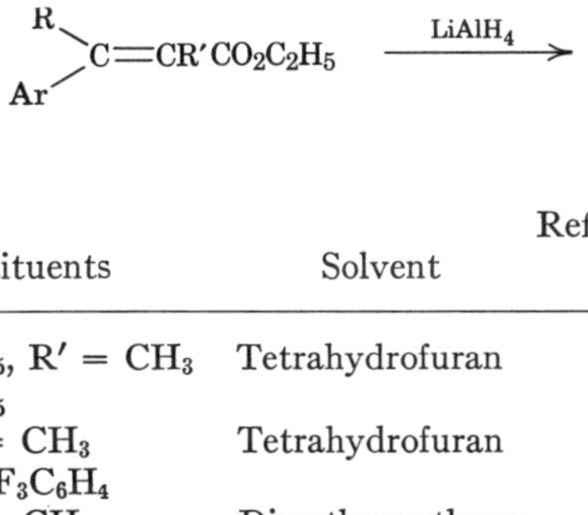

Substituents	Solvent	Reflux Time, Hours	Yield of Cyclopropane, %
R = C_6H_5, R′ = CH_3 Ar = C_6H_5	Tetrahydrofuran	10	78
R = R′ = CH_3 Ar = m-$CF_3C_6H_4$	Tetrahydrofuran	10	54
R = R′ = CH_3 Ar = C_6H_5	Dimethoxyethane	48	52
R = R′ = H Ar = C_6H_5[a]	Dimethoxyethane	320	66[b]
R = R′ = CH_3 Ar = 3,4-$Cl_2C_6H_3$	Tetrahydrofuran	1	45[b]

[a] Aldehyde rather than ester was employed; the yield was 28% after 3 days.

[b] From Reference 8.

1. Department of Chemistry, University of California, Berkeley, California 94720.
2. H. E. Simmons and R. D. Smith, *J. Am. Chem. Soc.*, **81**, 4256 (1959).
3. C. L. Bumgardner, *J. Am. Chem. Soc.*, **83**, 4420 (1961).
4. E. J. Corey and M. Chaykovsky, *J. Am. Chem. Soc.*, **87**, 1353 (1965).
5. H. Wieland and O. Probst, *Ann.*, **530**, 274 (1937).
6. W. R. Dunnavant and C. R. Hauser, *Org. Syntheses*, **44**, 56 (1964).
7. F. Bergmann, M. Weizmann, E. Dimant, J. Patai, and J. Szmuskowicz, *J. Am. Chem. Soc.*, **70**, 1612 (1948).
8. M. J. Jorgenson and A. W. Friend, *J. Am. Chem. Soc.*, **87**, 1815 (1965).
9. W. S. Wadsworth, Jr., and W. D. Emmons, *Org. Syntheses*, **45**, 44 (1965).

1,1-DIPHENYLPENTANE

(Pentane, 1,1-diphenyl-)

$$(C_6H_5)_2CH_2 \xrightarrow[NH_3]{NaNH_2} (C_6H_5)_2CHNa \xrightarrow[(C_2H_5)_2O]{n\text{-}C_4H_9Br} (C_6H_5)_2CHC_4H_9$$

Submitted by William S. Murphy, Phillip J. Hamrick, and Charles R. Hauser[1]
Checked by Prithipal Singh and Peter Yates

1. Procedure

Caution! This preparation should be carried out in a hood to avoid exposure to ammonia.

A suspension of sodium amide (0.275 mole) (Note 1) in liquid ammonia is prepared in the following manner in a 1-l. three-necked flask equipped with an air condenser (Note 2), a sealed mechanical stirrer, and a dropping funnel. Commercial anhydrous liquid ammonia (600 ml.) is introduced by pouring from an Erlenmeyer flask (Note 3). To the stirred liquid ammonia is added a small piece of sodium. After the appearance of a permanent blue color (Note 4) a few crystals of ferric nitrate hydrate (*ca.* 0.1 g.) are added, followed by small pieces of freshly cut sodium (Note 5) until 6.32 g. (0.275 g. atom) has been added. After all the sodium is converted to the amide (Note 6), 42.0 g. (0.250 mole) of diphenylmethane (Note 7) in 20 ml. of anhydrous ether is added (Note 8). The deep red suspension is stirred for 15 minutes. *n*-Butyl bromide (37.6 g., 0.274 mole) (Note 7) in

20 ml. of anhydrous ether is then added dropwise with stirring. The ammonia is allowed to evaporate (Note 9) from the resulting gray suspension. Water (100 ml.) is added carefully (Note 10), then 100 ml. of ether. The ethereal layer is separated, and the aqueous layer is extracted with two further 100-ml. portions of ether. The combined ethereal extracts are dried over Drierite and filtered, and the solvent is removed. The resulting liquid (54.5 g., 97%) is essentially pure 1,1-diphenylpentane (Notes 11 and 12). The liquid is distilled with the use of a Claisen distillation head without a fractionating column. The fraction, b.p. 138–139° (1.5 mm.), n^{26}_{D} 1.5501, weighs 51.6 g. (92%) (Note 13).

2. Notes

1. A 10% excess of sodium amide and *n*-butyl bromide with respect to diphenylmethane was adopted.

2. The checkers used a dry-ice condenser in place of the air condenser.

3. Dry commercial liquid ammonia is conveniently transferred from the cylinder via an Erlenmeyer flask without cooling and without the use of a condenser.

4. A permanent blue color may not remain after the addition of one pellet of sodium because of the presence of traces of moisture. Another pellet is added if necessary.

5. Sodium is cut into small pellets in the atmosphere but weighed under dry benzene or toluene.

6. Conversion is indicated by the discharge of the blue color (*ca.* 30 minutes). The addition of another portion of ferric nitrate hydrate will catalyze the conversion.

7. Freshly distilled diphenylmethane and *n*-butyl bromide were used.

8. The checkers found it important to add the diphenylmethane slowly (*ca.* 20 minutes); fast addition caused the reaction to get out of control.

9. The ammonia is allowed to evaporate overnight. A steam bath may be employed with care to facilitate the evaporation.

10. In the event of the presence of traces of unreacted sodium on the flask, water is added initially with special care.

11. The purity of the 1,1-diphenylpentane is attested by vapor-phase chromatography on a 5-ft. column of 10% Apiezon L on Celite at 200°.

12. Although 1,1-diphenylpentane undergoes air oxidation,[2] it appears to be stable in a stoppered flask under an inert atmosphere.

13. The checkers observed b.p. 127–129° (1.5 mm.).

3. Methods of Preparation

This procedure is an adaptation of one described by Hauser and Hamrick.[3] 1,1-Diphenylpentane has been prepared by the catalytic hydrogenation of 1,1-diphenyl-l-pentene[4-6] and in low yield from the reaction of diphenylmethyl bromide with di-*n*-butylmercury.[7] More recently 1,1-diphenylpentane was prepared by allowing lithium diphenylmethide to react with tri-*n*-butyl orthophosphate.[8]

4. Merits of the Preparation

This procedure illustrates a process which is general for 1,1-diphenyl substituted hydrocarbons. Diphenylmethane has been alkylated[3] with benzyl chloride, benzhydryl chloride, α-phenylethyl chloride, β-phenylethyl chloride, isopropyl chloride, 2-ethylbutyl bromide, and *n*-octyl bromide in yields of 99, 96, 97, 88, 86, 96, and 99%, respectively.

The present method is superior to earlier ones in that it is shorter, the chemicals are readily available, and high yields are obtained. The Gilman method[8] affords a 74% yield but a longer reaction time (1–2 days) and less readily available starting materials make it less convenient.

1. Department of Chemistry, Duke University, Durham, North Carolina 27706. This work was supported by the National Science Foundation.
2. M. S. Eventova, P. P. Borisov, and M. V. Chistyakova, *Vestn. Mosk. Univ. Ser. Mat., Mekhan., Astron., Fiz. i Khim.*, **12**, No. 3, 185 (1957) [*C.A.*, **52**, 6275 (1958)].
3. C. R. Hauser and P. J. Hamrick, Jr., *J. Am. Chem. Soc.*, **79**, 3142 (1957).
4. K. T. Serijan and P. H. Wise, *J. Am. Chem. Soc.*, **73**, 4766 (1951).
5. G. Benoit and F. Eliopoulo, *Bull. Soc. Chim. France*, [5] **18**, 892 (1951).

6. G. Wittig and E. Stahnecker, *Ann.*, **605**, 69 (1957).
7. F. C. Whitmore and E. N. Thurman, *J. Am. Chem. Soc.*, **51**, 1500 (1929).
8. H. Gilman and B. J. Gaj, *J. Am. Chem. Soc.*, **82**, 6326 (1960).

1-ETHYL-3-(3-DIMETHYLAMINO)PROPYLCARBODIIMIDE HYDROCHLORIDE AND METHIODIDE

(Carbodiimide, [3-(dimethylamino)propyl]ethyl-, hydrochloride and Ammonium iodide, [3-[[(ethylimino)methylene]amino]propyl]-trimethyl-)

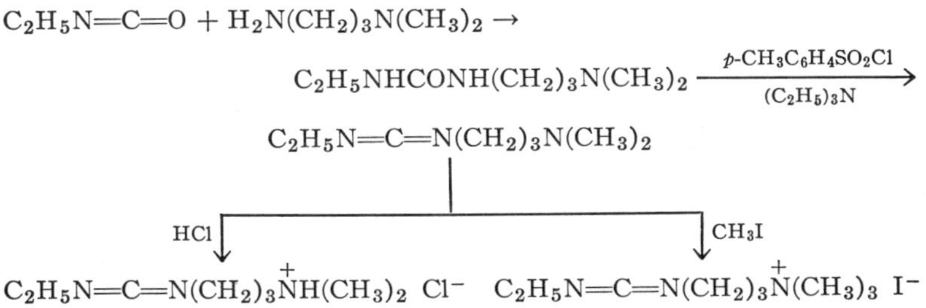

Submitted by John C. Sheehan and Philip A. Cruickshank [1]
Checked by E. J. Corey and Jerome E. Anderson

1. Procedure

A. *1-Ethyl-3-(3-dimethylamino)propylcarbodiimide.* A solution of 100 g. (1.41 moles) of ethyl isocyanate (Note 1) in 750 ml. of methylene chloride is prepared in a 5-l. three-necked flask equipped with a mechanical stirrer, an immersion thermometer, and a 500-ml., pressure-equalizing, addition funnel (Note 2). The flask and its contents are cooled to 5° in an ice bath, and a solution of 144 g. (1.41 moles) of N,N-dimethyl-1,3-propanediamine in 250 ml. of methylene chloride is added through the addition funnel at a rate such that the reaction temperature does not exceed 10° (Note 3). On completion of this addition 500 ml. of triethylamine is added to the flask, and a solution of 300 g. (1.6 moles) of *p*-toluenesulfonyl chloride in 300 ml. of methylene chloride is placed in the addition funnel and added to the reaction mixture, again at a rate such that the temperature does not exceed 10° (Note 4). After completion of the second addition, the ice bath is replaced with a heating mantle and the addition funnel with a reflux condenser; the reaction mixture is then heated under gentle reflux for 3 hours. Anhydrous sodium carbonate (400 g.) is added to the cooled reaction mixture, followed by 3.5 l. of ice water. The mixture is stirred vigorously for 30 minutes, after which the phases are allowed to separate, and the lower, organic phase is drawn off through a tube into a 2-l. suction flask. The aqueous phase is then extracted with three 500-ml. portions of methylene chloride (Note 5). The original methylene chloride phase and the extracts are combined and dried over anhydrous magnesium sulfate. The solvents are removed under reduced pressure, and the dark brown residue is distilled under reduced pressure through a 15-cm. Vigreux column to give 1-ethyl-3-(3-dimethylamino)propylcarbodiimide, b.p. 52–55° (0.3–0.4 mm.), n^{25}_{D} 1.4591. The yield is 110–118 g. (50–54%) (Note 6).

B. *1-Ethyl-3-(3-dimethylamino)propylcarbodiimide hydrochloride.* A suspension of 34.6 g. (0.300 mole) of pyridine hydrochloride (Note 7) in 280 ml. of methylene chloride is prepared in a 1-l. Erlenmeyer flask. To this is slowly added 46.5 g. (0.300 mole) of 1-ethyl-3-(3-dimethylamino)propylcarbodiimide. The resulting solution is diluted with anhydrous ether (Note 8) and stored at 0–5° for 16–20 hours. The crystalline product is collected by filtration in a *dry* atmosphere (Note 9), washed with a little anhydrous ether, and dried under reduced pressure over phosphorus pentoxide. The yield is 50.5–55.5 g. (88–96.5%), m.p. 104–109° (Notes 10 and 11). This material is sufficiently pure for most purposes.

C. *1-Ethyl-3-(3-dimethylamino)propylcarbodiimide methiodide.* A solution of 30.0 g. (0.193 mole) of 1-ethyl-3-(3-dimethylamino)-propylcarbodiimide in 750 ml. of anhydrous ether is prepared in a 2-l. Erlenmeyer flask. To this is slowly added from an addition funnel a solution of 30.0 g. (0.21 mole) of methyl iodide in 100 ml. of ether. The mixture is stored in the dark for 48 hours, after which the crystalline product is collected by filtration, washed with ether, and dried. The product, m.p. 94–95°, weighs 50.5–52.5 g. (88–91.5%) (Note 12).

2. Notes

1. Available from Eastman Organic Chemicals.
2. The assembled apparatus was dried at 120° for 18 hours.
3. The addition time was *ca.* 2 hours.
4. The addition time was *ca.* 3 hours.
5. Extractions were carried out in the reaction flask; after separation, the lower, methylene chloride phase was drawn off by suction.
6. The submitters, working on a twofold scale, obtained a yield of 60–65%.
7. Pyridine hydrochloride is extremely hygroscopic; the material used must be the anhydrous crystalline form.
8. Vigorous boiling of solvent occurs during addition of the carbodiimide and during spontaneous crystallization of the product. Approximately 250 ml. of ether was used.
9. The product is hygroscopic and care must be taken to protect it from atmospheric moisture at all times.
10. A sample of analytical purity had m.p. 113.5–114.5°.
11. The submitters, working on a 4.5-fold scale, obtained 92% of product, m.p. 108–112°.
12. The submitters, working on a twofold scale, obtained 95.5% of product, m.p. 93–95°, after one recrystallization from chloroform-ether.

3. Methods of Preparation

This procedure for the preparation of 1-ethyl-3-(3-dimethylamino)propylcarbodiimide and its salts is a modification of one that has been published.[2] Unsymmetrical carbodiimides have also been prepared by desulfurization of the corresponding thioureas with mercuric oxide [3] or by dehydration of the corresponding ureas with *p*-toluenesulfonyl chloride in pyridine.[4] Unsymmetrical 1,3-disubstituted ureas are best prepared by the reaction of isocyanates with primary or secondary amines[5] or by the action of carbamoyl chlorides on primary or secondary amines.[6]

4. Merits of the Preparation

Carbodiimides are, in general, useful compounds for effecting certain dehydrative condensations, *e.g.*, in the formation of amides, esters, and anhydrides. These two crystalline water-soluble carbodiimides are especially useful in the synthesis of peptides and in the modification of proteins. The excess of reagent and the co-product (the corresponding urea) are easily separated from products with limited solubility in water. The hydrochloride is best employed in nonaqueous solvents (methylene chloride, acetonitrile, dimethylformamide). The methiodide is relatively stable in neutral aqueous systems, and thus is recommended for those media.

Preparation of carbodiimides by dehydration of the corresponding ureas is of general applicability and is well adapted to the laboratory preparation of substantial quantities. The intermediates for this particular preparation are commercially available at moderate cost.

1. Research Institute for Medicine and Chemistry, Cambridge, Massachusetts 02142.
2. J. C. Sheehan, P. A. Cruickshank, and G. L. Boshart, *J. Org. Chem.*, **26**, 2525 (1961).
3. H. G. Khorana, *Chem. Rev.*, **53**, 145 (1953).
4. G. Amiard and R. Heymès, *Bull. Soc. Chim. France*, 1360 (1956).
5. V. Papesch and E. F. Schroeder, *J. Org. Chem.*, **16**, 1879 (1951).
6. E. N. Abrahart, *J. Chem. Soc.*, 1273 (1936).

2,3,4,5,6,6-HEXAMETHYL-2,4-CYCLOHEXADIEN-1-ONE

(2,4-Cyclohexadien-1-one, 2,3,4,5,6,6-hexamethyl-)

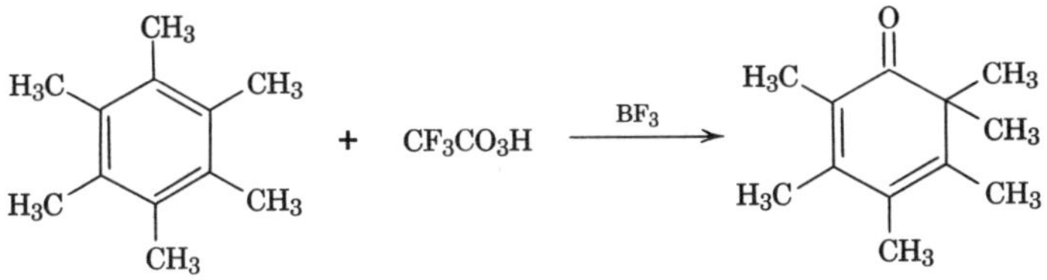

Submitted by HAROLD HART, RICHARD M. LANGE, and PETER M. COLLINS [1]
Checked by A. S. PAGANO and WILLIAM D. EMMONS

1. Procedure

Caution! The preparation and handling of peroxytrifluoroacetic acid should be carried out behind a safety screen. Precautions to be observed with 90% hydrogen peroxide are described in Note 3 and should be followed carefully.

To a 300-ml. three-necked flask equipped with a glass Trubore stirrer and two loose-fitting ground-glass stoppers are added 46.2 g. (0.22 mole) of trifluoroacetic anhydride and 50 ml. of methylene chloride (Note 1). The stirred solution is cooled in an ice bath, and 5.40 ml. (0.20 mole) of 90% hydrogen peroxide is added from a 10-ml. graduated cylinder in *ca.* 1-ml. portions over a period of 10 minutes (Notes 2, 3, and 4). When the mixture has become homogeneous, it is allowed to warm to room temperature for a few minutes and is then cooled once more in an ice bath to 0°.

A solution of 24.2 g. (0.15 mole) of hexamethylbenzene in 300 ml. of distilled methylene chloride is prepared in a 1-l. three-necked flask equipped with two ice-jacketed addition funnels (Note 5) and a thermometer. The solution is cooled to 5° in an ice-ethanol bath and is agitated by a magnetic stirrer. The cold peroxytrifluoroacetic acid solution is added at a constant rate to the hexamethylbenzene solution from one of the ice-jacketed addition funnels at the same time that 63.3 ml. of technical (48%) boron trifluoride etherate is added from the second addition funnel. The additions require *ca.* 45 minutes, and as far as possible they should be completed at the same time. During this period the temperature of the reaction mixture is maintained between 0° and 5° (Notes 6 and 7).

The mixture is stirred at 0–5° for 1 hour after addition is complete and then is hydrolyzed with 100 ml. of water, which is added quickly. The reaction mixture is extracted consecutively with two 100-ml. portions of water, three 100-ml. portions of saturated aqueous sodium bicarbonate, one 75-ml. portion of aqueous 5% sodium hydroxide, and two 75-ml. portions of water. The organic phase is dried over anhydrous magnesium sulfate, and the solvent is removed on a rotary evaporator. The residue, a mobile yellow oil, is distilled through a 6-in. Vigreux column under reduced pressure to give pure (Note 8) 2,3,4,5,6,6-hexamethyl-2,4-cyclohexadien-1-one, b.p. 85–87° (1.0 mm.). The yield is 22–24 g. (82–90%) (Note 9).

2. Notes

1. Excess anhydride is used to remove water introduced during the addition of 90% hydrogen peroxide.

2. Available from FMC Corp., Inorganic Chemicals Division, 808 Gwynne Building, Cincinnati 2, Ohio.

3. The precautions to be observed with 90% hydrogen peroxide have been described in detail.[2] In essence it is important to prevent contact of this reagent with any easily oxidizable substrates, such as wood, alcohols, and sugars and with heavy-metal salts, since the latter catalyze its decomposition. Storage of hydrogen peroxide in the laboratory should be arranged in such a way that, even if the bottle containing the reagent breaks, the hydrogen peroxide does not come into contact with any materials of this kind. Small samples of 90% hydrogen peroxide are regularly shipped in vented glass bottles provided with a protective outside metal container, and it is desirable to use this container while storing the reagent in the laboratory. If spillage of the reagent

occurs, dilution with at least several volumes of water is recommended. In weighing out 90% hydrogen peroxide it is good practice never to return excess reagent to the stock bottle; rather it should be diluted with water and discarded to avoid any possibility that the stock bottle will be contaminated.

4. The hydrogen peroxide may be added in one portion, but then an appreciable exotherm is noted.

5. A simple procedure for the construction of a jacketed addition funnel has been described.[3]

6. Boron trifluoride gas may be used in place of the etherate. In this case a fritted-glass gas-dispersion tube that extends below the liquid surface replaces the second addition funnel. Boron trifluoride gas (0.20 mole, 4.48 l.) is passed through the solution as the peroxytrifluoroacetic acid is added. The boron trifluoride may be metered into the mixture through a calibrated flowmeter containing carbon tetrachloride as the indicator liquid. Alternatively, a premeasured quantity of boron trifluoride may be displaced by carbon tetrachloride from a gas bulb. The yield is approximately the same regardless of the source of boron trifluoride.

7. If boron trifluoride is omitted as a reactant, the yield falls to about 67%.

8. This material should be at least 98% pure by vapor-phase chromatography (SE-30 column at 180–200°). It usually crystallizes if stored in a refrigerator. Unreacted hexamethylbenzene, present if insufficient oxidant is used, can best be removed by column chromatography on alumina with pentane as eluant.

9. The reaction can be used to prepare hexaethyl-2,4-cyclohexadienone, m.p. 44–45°, in 82% yield from hexaethylbenzene and 3,4,6,6-tetramethyl-2,4-cyclohexadienone from durene in over 80% yield.

3. Methods of Preparation

The method described is that of Waring and Hart.[4] Dienones of this type have not been available by any previously described synthetic route.

4. Merits of the Preparation

Dienones of this class are useful starting materials for the preparation of bicyclic compounds via Diels-Alder reactions[4] and for the synthesis of small ring compounds.[5] The 2,4-dienone can be converted quantitatively to the 2,5-isomer by treatment with fuming sulfuric acid and subsequent hydrolysis.[6] The oxidation procedure is also applicable to the conversion of mesitylene to mesitol or of isodurene to isodurenol,[7] and can be used to convert tetramethylethylene quantitatively and directly to pinacolone.[8]

1. Michigan State University, East Lansing, Michigan 48823.
2. E. S. Shanley and F. P. Greenspan, *Ind. Eng. Chem.*, **39**, 1536 (1947).
3. R. Graf, *Org. Syntheses*, **46**, 51 (1966).
4. A. J. Waring and H. Hart, *J. Am. Chem. Soc.*, **86**, 1454 (1964).
5. H. Hart and A. J. Waring, *Tetrahedron Letters*, 325 (1965).
6. H. Hart and D. W. Swatton, *J. Am. Chem. Soc.*, **89**, 1874 (1967).
7. H. Hart and C. A. Buehler, *J. Org. Chem.*, **29**, 2397 (1964).
8. H. Hart and L. R. Lerner, *J. Org. Chem.*, **32**, 2669 (1967).

2-HYDROXY-3-METHYLISOCARBOSTYRIL

(Isocarbostyril, 2-hydroxy-3-methyl-)

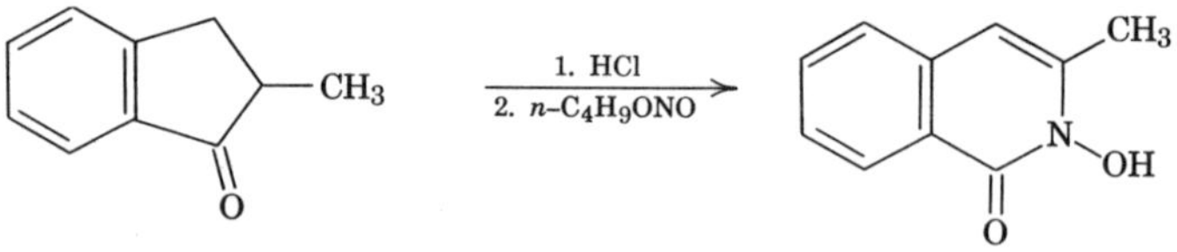

Submitted by Emil J. Moriconi and Francis J. Creegan[1]
Checked by Barbara A. Alexander, Hermann Ertl, T. Hoekemeijer, and Peter Yates

1. Procedure

A solution of 8.0 g. (7.5 ml., 0.055 mole) of 2-methyl-1-indanone (Note 1) in 100 ml. of toluene is prepared in a 500-ml., three-necked, round-bottomed flask equipped with a thermometer, a dropping funnel, and a magnetic stirrer. The flask is immersed in a freezing mixture of sodium chloride and ice. When the solution temperature reaches 0°, 70 ml. (0.21 mole) of 3*N* hydrochloric acid in ethyl acetate (Note 2) is added slowly (Note 3). To this mixture 8.0 ml. (0.068 mole) of freshly prepared *n*-butyl nitrite (Note 4) in 25 ml. of toluene is added with stirring over a 10-minute period (Note 5). The mixture is stirred for 1 hour at 0° and for an additional hour at room temperature. The two layers are separated, and the upper layer (toluene) is concentrated to one-half volume. Both solutions are refrigerated at −20° for 4 days. The precipitated orange product is collected by filtration from each layer. Further concentration of each filtrate under reduced pressure to one-half volume gives additional crude product (Note 6).

The various fractions are combined, washed with 20 ml. of cold ether, and dried. Recrystallization from methylene chloride-ether (Note 7) gives 6.0–6.6 g. (62–69%) of 2-hydroxy-3-methylisocarbostyril as light orange plates, m.p. 175–180°. Sublimation of this material at 100–110° (0.5 mm.) gives a white product, m.p. 182–184°, with softening at 174° (Note 7).

2. Notes

1. The 2-methyl-1-indanone, b.p. 65–66° (0.6 mm.) [lit.[2] b.p. 120° (15 mm.)], was prepared by the following method, described by Colonge and Weinstein.[2]

To 15.0 g. (0.50 mole) of paraformaldehyde (Eastman Organic Chemicals) and 100 g. (0.75 mole) of propiophenone (Eastman Organic Chemicals) in a 250-ml. Erlenmeyer flask, 10 ml. of 1*N* alcoholic potassium hydroxide solution was added with stirring. After a few minutes a clear solution formed, and the temperature rose to 35° and then fell slowly. The yellow solution was stirred for 5.5 hours at room temperature, during which time the solution became turbid. The turbid solution was poured into 150 ml. of water, and the mixture was acidified with concentrated hydrochloric acid (Congo red indicator).

The mixture was extracted with two 150-ml. portions of benzene, and the combined organic extracts were washed with two 150-ml. portions of water, two 150-ml. portions of 10% aqueous sodium carbonate, and two 150-ml. portions of water. The benzene extracts were dried over anhydrous sodium sulfate, and the solvent was removed. The yellow residue was distilled under reduced pressure to give a forerun consisting of 45 g. (0.34 mole) of unconsumed propiophenone followed by 32–36 g. (39–44%) of 3-hydroxy-2-methylpropiophenone, b.p. 108–110° (0.55 mm.) [lit.[2] b.p. 158–162° (17 mm.)]; infrared band (neat) at 5.94 μ (C=O).

3-Hydroxy-2-methylpropiophenone (30 g., 0.183 mole) was added slowly to 150 ml. of concentrated sulfuric acid with stirring. The temperature rose and the solution turned dark brown. The temperature remained at 80° for 10 minutes and then slowly fell. After 1 hour the dark solution was poured onto 200 g. of cracked ice. The mixture was extracted with two 100-ml. portions of ether. The ethereal solution was washed with two 100-ml. portions of water, two 100-ml. portions of saturated aqueous sodium bicarbonate, and again with two 100-ml. portions of water. It was dried over anhydrous potassium carbonate, and the solvent was removed. The residue was distilled to give 18–19 g. (67–71%) of a pale yellow liquid, b.p. 65–66° (0.6 mm.) [lit[2]. b.p. 120° (15 mm.)], n^{20}_{D} 1.5510 (lit.[2] n^{23}_{D} 1.5511); infrared band (neat) at 5.80 μ (C=O).

2. Prepared by dissolving 17.5 ml. of 12*N* hydrochloric acid in 52.5 ml. of ethyl acetate.

3. Two phases are obtained; this heterogenous mixture is vigorously stirred during the addition of *n*-butyl nitrite.

4. The *n*-butyl nitrite must be refrigerated after preparation [3] and used as soon as possible thereafter. The use of commercially available *n*-butyl nitrite invariably led to lower yields of the isocarbostyril.

5. With lower hydrochloric acid concentration and reversal of the mode of addition, *i.e.*, acid to indanone-nitrite mixture, the intermediate 2-methyl-2-nitroso-1-indanone may also be isolated as its dimer. This can be isomerized to the isocarbostyril rapidly in refluxing methanolic sodium methoxide and more slowly in concentrated hydrochloric acid.[4]

6. To determine whether all the isocarbostyril has been isolated from the filtrates, a small aliquot of the filtrate is treated with excess aqueous ferric chloride. The appearance of a deep purple

color indicates the necessity for further concentration under reduced pressure and precipitation of product.

7. The checkers used methylene chloride alone as the solvent for recrystallization; sublimation gave a product, m.p. 178–180°.

3. Methods of Preparation

2-Hydroxy-3-methylisocarbostyril has been prepared by the present method,[4] and in 12–15% yield by the ozonization of 3-methylisoquinoline-2-oxide.[5]

4. Merits of the Preparation

This simple, one-step ring expansion is the only available method for the preparation of 2-hydroxy-3-alkylisocarbostyrils in good yield from the corresponding 2-alkyl-1-indanones. Table I lists five new hydroxyisocarbostyrils prepared in this manner.

TABLE I

SYNTHESES OF 2-HYDROXY-3-ALKYLISOCARBOSTYRILS

2-Alkyl Substituent	Yield, %	M.P., °C
Ethyl	65	154–155
Propyl	64	139–141
Isopropyl	49	107–108
Butyl	45	108–109
t-Butyl	20	104–106

Direct reduction of the 2-hydroxy-3-alkylisocarbostyrils gives 3-alkylisocarbostyrils and provides a useful synthesis of these compounds.

1. Contribution No. 842 from the Department of Chemistry, Fordham University, New York, N. Y. 10458. This work was supported by the Directorate of Chemical Sciences, Air Force Office of Scientific Research, under Grant AF-AFOSR-62-18 and 488-64.

2. J. Colonge and G. Weinstein, *Bull. Soc. Chim. France*, [5] **19**, 462 (1952).
3. W. A. Noyes, *Org. Syntheses*, Coll. Vol. **2**, 108 (1943).
4. E. J. Moriconi, F. J. Creegan, C. K. Donovan, and F. A. Spano, *J. Org. Chem.*, **28**, 2215 (1963); E. J. Moriconi and F. J. Creegan, *J. Org. Chem.*, **31**, 2090 (1966).
5. E. J. Moriconi and F. A. Spano, *J. Am. Chem. Soc.*, **86**, 38 (1964).

3-HYDROXYPYRENE

(1-Pyrenol)

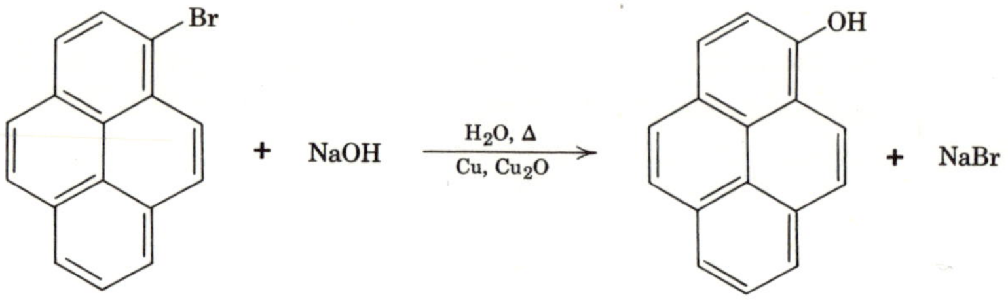

Submitted by W. H. GUMPRECHT[1]
Checked by WILLIAM G. DAUBEN, JOHN R. WISEMAN, and MICHAEL H. MCGANN

1. Procedure

Caution! Many pyrene derivatives are carcinogens. Contact of the skin with these materials should be avoided.

In a 100-ml. pressure vessel (Note 1) are placed 14.0 g. (0.050 mole) of 3-bromopyrene (Note 2), 0.5 g. of copper bronze powder (Note 3), 1.5 g. of cuprous oxide (Note 4), and 60 ml. of 10% aqueous sodium hydroxide (Note 5). The vessel is sealed, heated rapidly with shaking to 275–280°, and maintained at this temperature for 3 hours. The vessel is allowed to cool and is opened, and the contents are poured into a 500-ml. beaker containing 200 ml. of water (Note 6). The mixture is filtered, and the filter cake is washed with water until the washings become neutral to pH paper. The combined filtrate and washings, which show a blue fluorescence, are made acid to Congo red paper with 20% aqueous sulfuric acid. The precipitate is collected by filtration, washed free of acid with water, and dried in an oven at 100° to give *ca.* 9 g. of a gray solid.

The solid is boiled under reflux with *ca.* 175 ml. of benzene for 1 hour, and the benzene-insoluble material is removed from the hot mixture by filtration through a medium-grade sintered-glass funnel (Note 7). Small portions of activated alumina are added to the hot filtrate until a strong green fluorescence develops; 20–25 g. is needed (Note 8). The mixture is boiled (Note 9) under reflux for 15 minutes and filtered hot. On concentration and cooling, the fluorescent, lemon-yellow filtrate yields 3-hydroxypyrene as light yellow needles, m.p. 179–181°. An additional quantity of equally pure product is obtained by further concentration of the mother liquor. The total yield is 5.5–6.0 g. (50–55%) (Note 10).

2. Notes

1. Stainless steel and Hastelloy-C vessels were used with equivalent results.

2. This material, m.p. 93–95°, is readily prepared in high yields (78–86%) from commercial pyrene by the method of the submitter,[2] who used pyrene, m.p. 151–153°, obtained from Chemicals Division, Union Carbide Corp.

3. Copper bronze, type 3310, obtained from U. S. Bronze Powder Works, Inc., Flemington, New Jersey, was used. The use of some grades of copper powder leads to a considerably lower yield.

4. Technical grade cuprous oxide obtained from Baker and Adamson Products, General Chemical Division, Allied Chemical Corp., was used.

5. The amount of sodium hydroxide used does not affect the yield, provided that it is present in a quantity well in excess of that required by the stoichiometry.

6. The sodium salt of 3-hydroxypyrene is somewhat insoluble in the reaction mixture. Dilution before filtration ensures its removal from the copper residues.

7. The funnel should be preheated to prevent crystallization of the product in its pores. The filtration process can be accelerated by scraping the muddy cake from the funnel surface.

8. The use of a larger quantity reduces the yield with no significant improvement in quality.

9. The presence of the alumina causes the mixture to bump violently. Agitation of the boiling mixture with a magnetically

driven stirring bar helps to alleviate this problem. The vessel should be securely clamped.

10. The submitter obtained 25–29 g. (57–67%) of product when the reaction was run with 56.2 g. (0.20 mole) of the bromide in a 400-ml. pressure vessel.

3. Methods of Preparation

This procedure is based on the method of Smith, Opie, Wawzonek, and Prichard[3] for the preparation of 2,3,6-trimethylphenol. 3-Hydroxypyrene has been prepared by fusion of pyrene-3-sulfonic acid with sodium hydroxide[4] and by desulfonation of 3-hydroxypyrene-5,8,10-trisulfonic acid with hot, dilute sulfuric acid.[5]

4. Merits of the Preparation

It has been shown[6] that two mechanisms, elimination-addition (benzyne) and S_N2 displacement, are operative in the liquid-phase hydrolysis of halogenated aromatic compounds. The formation of isomeric phenols as a result of the availability of the benzyne route makes the reaction of limited synthetic value. The incorporation of the copper-cuprous oxide system suppresses reaction via the benzyne route, so that the present method has general utility for the preparation of isomer-free phenols. For example, *p*-cresol is the only cresol formed from *p*-bromotoluene under the conditions of this preparation.

The methods previously reported for the preparation of 3-hydroxypyrene have been found to be unsatisfactory, because of both very poor yields and difficulties in operation.

1. Contribution No. 300 from the Organic Chemicals Department, E. I. du Pont de Nemours and Company, Wilmington, Delaware 19899.
2. W. H. Gumprecht, *Org. Syntheses*, this volume, p. 30.
3. L. I. Smith, J. W. Opie, S. Wawzonek, and W. W. Prichard, *J. Org. Chem.*, **4**, 318 (1939).
4. H. Vollmann, H. Becker, M. Corell, and H. Streeck, *Ann.*, **531**, 1 (1937); W. Kern, U. S. Patent 2,018,792 (1935) [*C.A.*, **30**, 112 (1936)].
5. E. Tietze and O. Bayer, *Ann.*, **540**, 189 (1939).
6. A. T. Bottini and J. D. Roberts, *J. Am. Chem. Soc.*, **79**, 1458 (1957).

2-METHYLMERCAPTO-N-METHYL-Δ^2-PYRROLINE

(2-Pyrroline, 1-methyl-2-methylthio-)

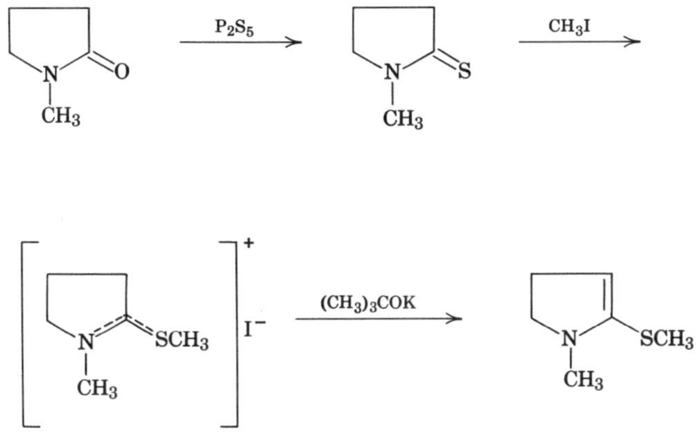

Submitted by R. GOMPPER and W. ELSER [1]
Checked by HERMANN ERTL, IAN D. RAE, and PETER YATES

1. Procedure

Caution! Hydrogen sulfide is very poisonous. Procedure A should be conducted in a hood.

A. *N-Methyl-2-pyrrolidinethione.* Phosphorus pentasulfide (667 g., 3.00 moles) is suspended in 600 ml. of carbon disulfide in a 2-l. three-necked flask equipped with a mechanical stirrer with a segment-shaped paddle, an efficient reflux condenser, and a dropping funnel. N-Methyl-2-pyrrolidinone (300 g., 3.03 moles) is added in large portions from the dropping funnel with vigorous stirring. The reaction mixture warms up considerably and, after addition of about two-thirds of the amide, becomes so viscous that further stirring is impossible. The addition of the remainder is carried out without stirring at a rate such that the solution boils gently. The semisolid, yellow-brown mixture is boiled under reflux on a water bath for a further 10 hours.

The carbon disulfide is decanted, and 200–300 ml. of water is added to the contents of the flask. Initially the mixture reacts slowly, but after some time the reaction becomes so vigorous that it is necessary to pour off the water (Notes 1 and 2). The addition of water, followed by its removal when the reaction becomes very vigorous, is repeated until decomposition is complete. The combined aqueous solutions are extracted several times with chloroform, and the combined extracts are dried over anhydrous sodium sulfate. Distillation of the solvent gives a dark brown liquid residue which is distilled under reduced pressure to give 310.3 g. (89%) of the thioamide as a yellow liquid, b.p. 133–135° (12 mm.) (Note 3).

B. *2-Methylmercapto-N-methyl-Δ^1-pyrrolinium iodide.* N-Methyl-2-pyrrolidinethione (310 g., 2.69 moles) is dissolved with stirring in 1.1 l. of anhydrous ether in a 2-l. three-necked flask equipped with a mechanical stirrer, a reflux condenser fitted with a drying tube, and a dropping funnel. To this solution is added *ca.* 5 g. of the product as seed crystals (Note 4) to prevent initial deposition of the iodide as an oil that suddenly crystallizes with considerable evolution of heat. Methyl iodide (520 g., 3.66 moles) is then added rather rapidly. The solution becomes turbid after a short time, and separation of the salt begins with heat evolution. After 12 hours the hygroscopic, crystalline paste is filtered and dried in a desiccator; yield 663 g. (96%).

This product can be used without further purification. Crystallization from mixtures of acetonitrile and ether gives colorless, felted needles, m.p. 118–120° (dec.) (Note 5).

C. *2-Methylmercapto-N-methyl-Δ^2-pyrroline.* 2-Methylmercapto-N-methyl-Δ^1-pyrrolinium iodide (662 g., 2.57 moles) is suspended in 1.25 l. of anhydrous ether in a 4-l. three-necked flask equipped with a mechanical stirrer and a reflux condenser with a segment-shaped paddle. Potassium *t*-butoxide (448 g., 4.0 moles) is added in one batch to this suspension with vigorous stirring. The mixture warms up a little, and later the solid becomes fine-grained and more mobile as a result of separation of potassium iodide. After being stirred for 1.5 hours at room temperature, the mixture is treated with 1.8 l. of anhydrous ether and boiled under reflux on a water bath for 5 hours.

The flask is cooled in an ice bath, and the precipitate is filtered onto a large Buchner funnel and washed well with ether (Note 6). Ether is distilled from the yellow-brown filtrate at atmospheric pressure, and *t*-butyl alcohol is distilled under reduced pressure at 40°. Distillation of the residue through a 10-cm. Vigreux column at the water aspirator gives, after a small forerun (*ca.* 20 ml.), the product as a colorless liquid, b.p. 70–73° (10–12 mm.); yield 268 g. (81%) (Notes 7, 8, and 9). This becomes brown on standing.

Repeated fractional distillation gives a product of analytical purity (Note 10), b.p. 65–68° (8–10 mm.), n^{25}_D 1.5222 (Note 11); yield 62%. This is an extremely disagreeable, musty liquid which on standing in the atmosphere warms up slightly and immediately turns red.

2. Notes

1. The large amount of hydrogen sulfide produced is destroyed by passage into two wash bottles containing concentrated aqueous potassium hydroxide.

2. It is also possible to moderate the reaction by the addition of chloroform.

3. The checkers, working at quarter scale, obtained 84–87% of product, b.p. 131–133° (11 mm.), 139–142° (14–15 mm.).

4. This can be prepared readily in a test tube by addition of methyl iodide to a solution of the thioamide in ether.

5. *Anal.* Calcd. for $C_6H_{12}INS$: I, 49.36. Found: 49.34.

6. The precipitate may be so fine-grained that the filtrate is initially turbid; the turbid filtrate is then filtered again through the Buchner funnel containing the precipitate.

7. This fraction is followed by tailings (12 g.), b.p. 74–75° (10–12 mm.).

8. The checkers, working at quarter scale, obtained 75% of product, b.p. 68–71° (13 mm.), followed by tailings, b.p. 71–74° (13 mm.).

9. This product is contaminated with N-methyl-2-pyrrolidinone, which does not impair its usefulness for most further reactions.

10. *Anal.* Calcd. for $C_6H_{11}NS$: C, 55.76; H, 8.58; N, 10.84; S, 24.82. Found: C, 55.60; H, 9.08; N, 10.84, S, 24.66.

11. The checkers found that the refractive index of the pyrroline increases rapidly on exposure to air, impairing its usefulness as a criterion of purity.

3. Methods of Preparation

The method described for the preparation of N-methyl-2-pyrrolidinethione is very similar to that of Peak and Stansfield [2] for the preparation of 4-thioacetylmorpholine. N-Methyl-2-pyrrolidinethione has also been prepared by the reaction of 2-chloro-N-methyl-Δ^1-pyrrolinium chloride with hydrogen sulfide [yield 83%; b.p. 144–145° (15 mm.)] [3] and by heating N-methyl-2-pyrrolidinone with 2 equivalents of phosphorus pentasulfide in xylene [yield not reported; b.p. 125–132° (10 mm.)].[4] General procedures for the preparation of N,N-disubstituted thioamides have been reviewed.[5, 6]

The method described for the preparation of 2-methylmercapto-N-methyl-Δ^1-pyrrolinium iodide is based on the general procedure of several authors.[2, 7–9] Its preparation has not been described previously; the corresponding methyl sulfate has been obtained as a noncrystalline, viscous mass.[4, 10]

2-Methylmercapto-N-methyl-Δ^2-pyrroline has been prepared by the present method only.[11, 12]

4. Merits of the Preparation

The procedure illustrates a general method for the preparation of ketene S,N-acetals via thioamides and their crystalline quaternary iodides; some other examples are shown in Table I. 2,2-Dialkyl-substituted ketene S,N-acetals cannot be prepared by this method because the nature of the products makes rigorous purification impractical. Ketene S,N-acetals are useful starting materials for many syntheses.[11, 13-15]

TABLE I

PREPARATION OF KETENE S,N-ACETALS

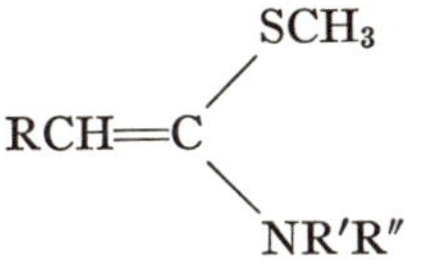

R	R′	R″	B.P., °C.	Yield of Crude Product, % (Purity, %)[a]
H	CH_3	CH_3	31–32 (11 mm.)	77 (92–96)[b]
H	$(CH_2)_4$		79–81 (10 mm.)	60 (72–88)
CH_3	CH_3	CH_3	38–40 (10 mm.)	68 (88)
C_2H_5	CH_3	CH_3	56 (11 mm.)	74 (92)
C_6H_5	$(CH_2)_2O(CH_2)_2$[c]		114–120 (10^{-3} mm.)[d]	81
$(CH_2)_3$		CH_3	75–76 (12 mm.)	69–73 (94)
$(CH_2)_4$		CH_3	86–87.5 (12 mm.)	82 (94)

[a] Determined by gas chromatography; products distilled through a spinning-band column were 94–99% pure.
[b] Yield of pure product, 52%.
[c] M. A. T. Rogers, *J. Chem. Soc.*, 3350 (1950).
[d] M.p. 44–45° (ex ethanol).

1. Institut für Organische Chemie der Universität München, München, Germany.
2. D. A. Peak and F. Stansfield, *J. Chem. Soc.*, 4067 (1952).
3. H. Eilingsfeld, M. Seefelder, and H. Weidinger, *Ber.*, **96**, 2671 (1963).
4. O.Riester and F. Bauer, German Patent 927,043 (1955) [*C.A.*, **52**, 938 (1958)].
5. A. Schöberl and A. Wagner, in E. Müller, "Methoden der Organischen Chemie (Houben-Weyl)," 4th ed., Vol. 9, Georg Thieme Verlag, Stuttgart, 1955, p. 762.
6. R. N. Hurd and G. DeLamater, *Chem. Rev.*, **61**, 45 (1961).
7. J. Renault, *Compt. Rend.*, **233**, 182 (1951).
8. P. Chabrier and S. H. Renard, *Compt. Rend.*, **228**, 850 (1949).
9. R. I. Meltzer, A. D. Lewis, and J. A. King, *J. Am. Chem. Soc.*, **77**, 4062 (1955).
10. S. Hünig and F. Müller, *Ann.*, **651**, 89 (1962).
11. W. Elser, Ph.D., Dissertation, Technische Hochschule, Stuttgart, 1965.
12. R. Gompper and W. Elser, *Tetrahedron Letters*, 1971 (1964).
13. R. Buijle, A. Halleux, and H. G. Viehe, *Angew. Chem.*, **78**, 593 (1966); *Angew. Chem. Intern. Ed. Engl.*, **5**, 584 (1966).
14. R. Gompper and W. Elser, *Angew. Chem.*, **79**, 382 (1967); *Angew. Chem. Intern. Ed. Engl.*, **6**, 366 (1967).
15. R. Gompper, W. Elser, and H.-J. Müller, *Angew. Chem.*, **79**, 473 (1967); *Angew. Chem. Intern. Ed. Engl.*, **6**, 453 (1967).

1-METHYL-3-*p*-TOLYLTRIAZENE

(Triazene, 1-methyl-3-*p*-tolyl-)

AND ITS USE IN THE ESTERIFICATION OF ACIDS

$$p\text{-}CH_3C_6H_4NH_2 + HONO \xrightarrow[HCl]{H_2O} p\text{-}CH_3C_6H_4N_2^+\ Cl^-$$

$$p\text{-}CH_3C_6H_4N_2^+\ Cl^- + CH_3NH_2 \xrightarrow{H_2O} p\text{-}CH_3C_6H_4N{=}NNHCH_3$$

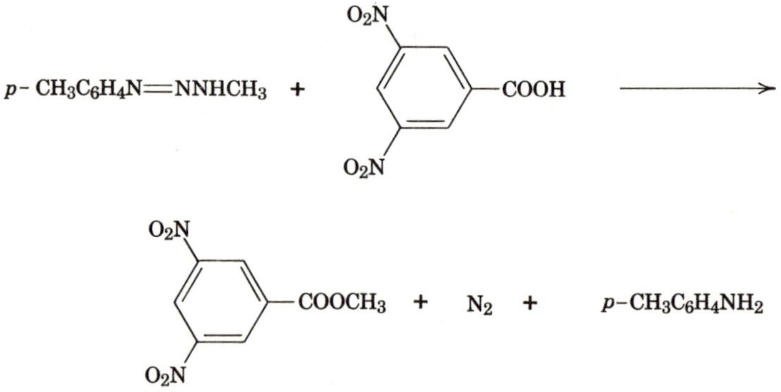

Submitted by E. H. White, A. A. Baum, and D. E. Eitel [1]
Checked by I. Katz and R. Breslow

1. Procedure

A. *1-Methyl-3-p-tolyltriazene*. *p*-Toluidine (50.2 g., 0.47 mole) is added to a 2-l. flask equipped with a 200-ml. dropping funnel and an efficient stirrer, and the flask is immersed in an ice-salt bath at *ca.* −10°. A solution of 46.8 g. (0.55 mole) of potassium nitrite in 150 ml. of water is placed in the dropping funnel, and a mixture of 250 g. of crushed ice and 140 ml. of concentrated hydrochloric acid is added to the *p*-toluidine with stirring. The potassium nitrite solution is slowly added with continued stirring during 1–2 hours until a positive starch-potassium iodide test is obtained (Note 1), and the mixture is stirred for an additional hour to ensure the reaction of all the toluidine.

The solution of *p*-toluenediazonium chloride is then brought to pH 6.8–7.2 at 0° with cold, concentrated, aqueous sodium carbonate, whereupon the solution becomes red to orange in color and a small amount of red material settles out. The cold, neutral solution is transferred to a dropping funnel and added slowly to a vigorously stirred mixture of 150 g. of sodium carbonate, 300 ml. of 30–35% aqueous methylamine (Note 2), and 100 g. of crushed ice in a 3-l. flask. The reaction mixture is kept at *ca.* −10° during the addition, which requires about 45 minutes (Note 3). The solution is extracted with three 1-l. portions of ether. The ethereal extracts are dried with anhydrous sodium sulfate and evaporated on a rotary evaporator at room temperature to give 65 g. of crude 1-methyl-3-*p*-tolyltriazene (Note 4). This is placed in a water-cooled sublimer, and the triazene is sublimed at 50° (1 mm.); 43.3 g. (0.29 mole, 62%) of a yellow, crystalline sublimate, m.p. 77–80°, is obtained (Note 5). The sublimate can be recrystallized from hexane to give the triazene as white needles, m.p. 80.5–81.5°. More conveniently, it is dissolved in the minimum amount of ether, and the solution is diluted with 2 volumes of hexane and cooled to 0° to give flat plates with a slightly yellow cast; m.p. 79–81°. The yield of pure triazene is 33–37 g. (47–53%) (Note 6).

B. *Esterification of 3,5-dinitrobenzoic acid with 1-methyl-3-p-tolyltriazene* (Note 7). A solution of 1-methyl-3-*p*-tolyltriazene (1.05 g., 7.0 mmoles) in 10 ml. of ether is placed in a 100-ml. flask equipped with a 100-ml. dropping funnel (Note 8). A solution of 1.50 g. (7.1 mmoles) of 3,5-dinitrobenzoic acid in 25 ml. of ether is placed in the dropping funnel and is slowly added to the triazene solution; the contents of the flask are gently swirled from time to time. During the addition, nitrogen is evolved, and the solution becomes red in color. After the nitrogen evolution has ceased (*ca.* 1 hour), the ethereal solution is transferred to a separatory funnel and washed with 5*N* hydrochloric acid to remove toluidine (Note 9). It is then washed with 5% aqueous sodium carbonate and dried over anhydrous sodium sulfate. Evaporation of the ether yields methyl 3,5-dinitrobenzoate (1.11–

1.42 g., 70–90%) as light tan crystals, m.p. 106–107.5° (Note 10). Recrystallization from ether yields small, flat plates, m.p. 107–107.5°.

2. Notes

1. The individual tests with starch-potassium iodide paper should be made 1–2 minutes after the addition of potassium nitrite has been stopped.

2. The checkers used 40% aqueous methylamine supplied by Matheson, Coleman and Bell.

3. The reaction is over when a drop of solution no longer gives a red color with a solution of β-naphthol in aqueous sodium carbonate.

4. The chief impurity is 1,5-di-*p*-tolyl-3-methyl-1,4-pentazadiene (m.p. 148°). This can be removed by fractional crystallization, but it is easier to sublime the triazene from the reaction mixture.

5. The sublimate contains a trace of 1,3-di-*p*-tolyltriazene, as shown by thin-layer chromatography. Recrystallization yields the pure 1-methyl-3-*p*-tolyltriazene.

6. This procedure works well only with water-soluble amines. A procedure has been given elsewhere for the preparation of triazenes of water-insoluble amines.[2]

7. The ethyl, propyl, butyl, and other esters may be prepared similarly from the corresponding triazenes.

8. Many solvents may be used for this reaction; the reaction rate, however, is greater in nonpolar solvents. Less color develops in the reaction mixture if the system is flushed with nitrogen at this point.

9. The colored impurities normally enter the aqueous phase.

10. The infrared spectrum of this material is essentially identical with that of the pure, recrystallized ester.

3. Methods of Preparation

1-Methyl-3-*p*-tolyltriazene has been prepared by the reaction of methylmagnesium bromide with *p*-tolyl azide,[3] and, in unspecified yield, by the addition of methylamine to *p*-toluenediazonium chloride.[4]

4. Merits of the Preparation

This procedure represents the most convenient synthesis of 1-methyl-3-*p*-tolyltriazene. Triazenes with more complex alkyl groups may be prepared from the corresponding amine[2] or Grignard reagent.[3]

The alkylation of acids with triazenes is superior to alkylation with diazomethane and other diazoalkanes in that the triazenes are crystalline, stable materials which are easy to prepare and store. Alkylations with triazenes are unlikely to be accompanied by side reactions, such as addition to strained or conjugated double bonds, which are frequently observed in alkylations with diazoalkanes.

1. Department of Chemistry, The Johns Hopkins University, Baltimore, Maryland 21218.
2. E. H. White and H. Scherrer, *Tetrahedron Letters*, 758 (1961).
3. O. Dimroth, M. Eble, and W. Gruhl, *Ber.*, **40**, 2390 (1907).
4. C. S. Rondestvedt, Jr., and S. J. Davis, *J. Org. Chem.*, **22**, 200 (1957).

D-2-OXO-7,7-DIMETHYL-1-VINYLBICYCLO[2.2.1]HEPTANE

(2-Norbornanone, 7,7-dimethyl-1-vinyl-, D-)

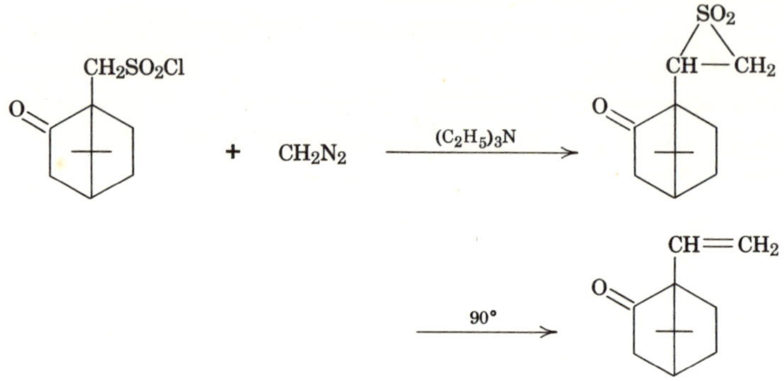

Submitted by NIKOLAUS FISCHER and G. OPITZ[1]
Checked by HERMANN ERTL, IAN D. RAE, and PETER YATES

1. Procedure

Caution! Diazomethane is both explosive and poisonous, and all operations involving its preparation and use must be carried out in a hood. Follow the directions for its handling given in earlier volumes.[2, 3]

In a 500-ml. three-necked flask equipped with a mechanical stirrer, a dropping funnel, and a reflux condenser fitted with a potassium hydroxide drying tube are placed 7.0 g. (0.069 mole) of triethylamine (Note 1) and a solution of 3.15 g. (0.075 mole) of diazomethane in 200 ml. of ether (Note 2). The flask is cooled in an ice bath, and a solution of 13.0 g. (0.052 mole) of D-camphor-10-sulfonyl chloride (Note 3) in 75 ml. of anhydrous ether is added dropwise over a period of 1 hour. Triethylamine hydrochloride slowly precipitates. The reaction mixture is stirred for an additional 30 minutes and then concentrated to *ca.* 150 ml. under reduced pressure (water aspirator) with continued stirring to remove the excess of diazomethane. The mixture is filtered under reduced pressure; the precipitate is washed with 50 ml. of anhydrous ether, giving 6.7 g. (94%) of triethylamine hydrochloride. The combined filtrate and washings are freed of solvent on a rotary evaporator at room temperature to give 10.7 g. (90%) of crude episulfone, m.p. 76–85° (dec.) (Note 4). This is used without purification in the next step; it can be purified by crystallization from a little methanol at −20°. This gives colorless episulfone, m.p. 83–85° (dec.), $[\alpha]^{24}\text{D}$ −6.72° (methanol, c = 3.20); infrared bands at 3070, 1300, and 1170 cm.$^{-1}$ (Note 5).

The crude episulfone (3.0 g.) is placed in a 10-ml. round-bottomed flask fitted with a reflux condenser and is heated at 95° for 30 minutes, when it decomposes with loss of sulfur dioxide. The reflux condenser is replaced with a distillation head (Note 6), and the yellow residue is distilled under reduced pressure (water aspirator). D-2-Oxo-7,7-dimethyl-1-vinylbicyclo[2.2.1]heptane (1.7 g., 71% based on sulfonyl chloride), b.p. 95–96° (10 mm.), distills at a bath temperature of 110–120°. Sublimation at 60° (0.01 mm.) gives the olefin as colorless, waxy crystals, m.p. 64–65°, $[\alpha]^{25}\text{D}$ +16.35° (methanol, c = 2.16); infrared band at 1650 cm.$^{-1}$.

2. Notes

1. The triethylamine was purified by treatment with naphthyl isocyanate and distilled; the distillate was stored over sodium wire.

2. The ethereal diazomethane was prepared from N-nitrosomethylurea and aqueous potassium hydroxide and dried over potassium hydroxide pellets for 2–3 hours. The solid potassium hydroxide was replaced once or twice to ensure complete dryness. The checkers used the procedure of Arndt [4] for this preparation and for the estimation of the diazomethane.

3. D-Camphor-10-sulfonyl chloride can be prepared from commercially available D-camphor-10-sulfonic acid and phosphorus pentachloride [5] or thionyl chloride.[6] The checkers used the following procedure. D-Camphor-10-sulfonic acid (50.0 g.) was added slowly to 50 g. of thionyl chloride. The mixture was boiled under reflux until homogeneous and then for a further 2 hours. The solution was cooled and poured onto *ca.* 500 g. of

crushed ice. The crude product (51.1 g., 95%) was filtered and crystallized twice from hexane to give the sulfonyl chloride, m.p. 65–67.5°; yield, 35.5 g. (66%).

4. The checkers obtained higher yields (94–97%) of less pure material [m.p. 50–78° (dec.)].

5. The checkers observed that the episulfone decomposed slowly at room temperature and, on one occasion, during evaporation at 15°.

6. The checkers found it advisable to use an apparatus with a wide-bore side arm (18 mm.) without a condenser.

3. Methods of Preparation

The only method reported for the preparation of D-2-oxo-7,7-dimethyl-1-vinylbicyclo[2.2.1]heptane is that of the present procedure.[7]

4. Merits of the Preparation

The method is of general applicability [7] for the synthesis of olefins. Other sulfonyl chlorides, RCH_2SO_2Cl, have been used where R = H, C_2H_5, C_6H_5, and $C_6H_5CH_2$; other diazoalkanes that have been used are diazoethane and 1-diazo-2-methylpropane. In all cases the olefins form without double-bond migration.

1. Chemisches Institut der Universität, Tübingen, Germany.
2. T. J. de Boer and H. J. Backer, *Org. Syntheses*, Coll. Vol. **4**, 250 (1963).
3. J. A. Moore and D. E. Reed, *Org. Syntheses*, **41**, 16 (1961).
4. F. Arndt, *Org. Syntheses*, Coll. Vol. **2**, 165 (1943).
5. P. D. Bartlett and L. H. Knox, *Org. Syntheses*, **45**, 14 (1965).
6. S. Smiles and T. P. Hilditch, *J. Chem. Soc.*, **91**, 519 (1907).
7. G. Opitz and K. Fischer, *Angew. Chem.*, **77**, 41 (1965); *Angew. Chem. Intern. Ed. Engl.*, **4**, 70 (1965) [Correction: Compounds (**1f**) and (**2i**) were obtained from 1-diazo-2-methylpropane (R^2 = H, R^3 = $CH(CH_3)_2$) and not from 2-diazobutane (R^2 = CH_3, R^3 = C_2H_5)].

PHENYLGLYOXAL

(Glyoxal, phenyl-)

$$CH_3SOCH_3 \xrightarrow{(CH_3)_3COK} CH_3SOCH_2^-K^+$$

$$C_6H_5CO_2C_2H_5 + CH_3SOCH_2^-K^+ \longrightarrow [C_6H_5COCHSOCH_3]^-K^+$$

$$\xrightarrow{H_3O^+} C_6H_5CO\overset{\overset{\large OH}{|}}{C}HSCH_3 \xrightarrow{Cu(OAc)_2} C_6H_5COCHO$$

Submitted by Gerard J. Mikol and Glen A. Russell[1]
Checked by William G. Dauben, Michael H. McGann, and Noel Vietmeyer

1. Procedure

A. *Phenylglyoxal hemimercaptal.* In a 1-l. three-necked flask equipped with an all-glass mechanical stirrer, a 125-ml. dropping funnel, and a condenser fitted with a nitrogen-inlet tube are placed 90 ml. (99 g., 1.27 moles) of dry dimethyl sulfoxide (Note 1), 120 ml. of dry *t*-butyl alcohol (Note 1), and 57.4 g. (0.51 mole) of potassium *t*-butoxide (Notes 2 and 3). The mixture is warmed to 80°; when all the solid has dissolved, the heating is discontinued, and 72 ml. (75 g., 0.50 mole) of dry ethyl benzoate (Note 1) is added slowly from the dropping funnel. The reaction mixture is stirred at room temperature for 4 hours, and the solvent is removed at 80–90° under reduced pressure until the volume of the reaction mixture has been reduced to 150 ml. (Note 4). The residue is poured into 500 ml. of an ice-water slurry. The resulting aqueous solution is extracted with three 100-ml. portions of ether, and the ethereal extracts are discarded (Note 5). The aqueous solution is acidified with a solution of 190 ml. of concentrated hydrochloric acid in 675 ml. of water, and the mixture is allowed to stand at room temperature for 30 hours. The pale yellow precipitate is removed by suction filtration, washed with 500 ml. of cold water, and air-dried to yield 69–74 g. (76–81%) of phenylglyoxal hemimercaptal, m.p. 103–105°.

B. *Phenylglyoxal.* The phenylglyoxal hemimercaptal prepared as described in procedure A (69–74 g.) is dissolved in 400 ml. of warm chloroform, and 60 g. (0.30 mole) of powdered cupric acetate monohydrate is added in one portion to the well-stirred solution. The mixture is stirred at room temperature for 1 hour; the solids are removed by suction filtration and washed with two 75-ml. portions of chloroform. The combined chloroform filtrate and washings are shaken in a separatory funnel with 75 ml. of water; 20 g. of powdered sodium carbonate is added in small portions to the funnel, and the chloroform solution is shaken with the neutralized aqueous solution. (*Caution! Carbon dioxide is evolved.*) The aqueous layer is separated and extracted with four 30-ml. portions of chloroform. The chloroform solutions are combined and dried with anhydrous magnesium sulfate, and the chloroform is removed under reduced pressure. The residue is fractionally distilled under reduced pressure to yield 43–49 g. (64–73%, based on ethyl benzoate) of anhydrous phenylglyoxal as a yellow liquid, b.p. 63–65° (0.5 mm.).

2. Notes

1. The presence of water results in very rapid saponification of ethyl benzoate. Dimethyl sulfoxide (Crown Zellerbach Corp.) may be dried by stirring with calcium hydride for 4–8 hours, followed by distillation under reduced pressure at 80–90° without filtration. Commercial *t*-butyl alcohol and ethyl benzoate are conveniently dried by stirring for 2–4 hours with calcium hydride followed by filtration.

2. Potassium *t*-butoxide was obtained from Mine Safety Appliances Corp.

3. The potassium salt of dimethyl sulfoxide can also be prepared in the following manner. In a 1-l. three-necked flask equipped with an all-glass mechanical stirrer, a 125-ml. dropping funnel containing 90 ml. of dry dimethyl sulfoxide (Note 1), and a Claisen distillation head and condenser is placed 425 ml. of dry *t*-butyl alcohol (Note 1). The system is flushed with dry

nitrogen, and 20 g. (0.51 g. atom) of potassium is added (Note 6). The system is closed to the atmosphere by a mineral oil bubbler through which the evolved hydrogen escapes. The mixture is stirred at 80° until the potassium has dissolved. After cooling, the unreacted alcohol is removed by distillation under reduced pressure until a thick slurry of potassium *t*-butoxide remains (Note 7). The dimethyl sulfoxide is added from the dropping funnel, and the mixture is heated to 80–90° to dissolve all the solid. The solution is maintained at this temperature, and additional *t*-butyl alcohol is removed under reduced pressure until the volume of the solution is reduced to 300 ml.

4. Since the volume of the solution at this point is critical, the reaction flask should be calibrated.

5. The aqueous solution can be used to prepare 2-(methylsulfinyl)acetophenone by the following procedure. The solution is acidified to pH 1–2 (Hydrion paper) by the slow addition of concentrated hydrochloric acid with vigorous stirring and is extracted immediately with two 100-ml. portions of chloroform. The chloroform extracts are combined, washed with 75 ml. of saturated aqueous sodium carbonate and two 75-ml. portions of water, and dried over anhydrous magnesium sulfate. The chloroform is removed under reduced pressure, and the resulting solid is pulverized, slurried with 100 ml. of ether, collected by filtration, and air-dried. The 2-(methylsulfinyl)acetophenone weighs 75–77 g. (82–85%); m.p. 85–86°. It can be converted to phenylglyoxal hemimercaptal by treatment with dilute hydrochloric acid in dimethyl sulfoxide solution at room temperature (2 ml. of dimethyl sulfoxide, 2 ml. of concentrated hydrochloric acid, and 15 ml. of water per gram of the keto sulfoxide). The solution is allowed to stand at room temperature for 30 hours, after which the phenylglyoxal hemimercaptal can be isolated as described in procedure A.

6. The potassium should be free of oxide and/or hydroxide to avoid subsequent saponification of ethyl benzoate.

7. A heating mantle may be used, but care must be taken to avoid decomposition on the walls of the flask due to overheating during the later stages of the distillation.

3. Methods of Preparation

Phenylglyoxal has been prepared from isonitrosoacetophenone via the bisulfite compound [2] and by treatment with nitrosyl sulfuric acid [3] or nitrous acid.[4] It has also been prepared by oxidation of benzoylcarbinol with cupric acetate,[5] by heating or aqueous hydrolysis of 2-acetoxy-2-bromoacetophenone,[6] by selenium dioxide oxidation of acetophenone,[7] by oxidation of phenacyl bromide with dimethyl sulfoxide,[8] by oxidative bromination of phenylglyoxal diethyl mercaptal,[9] and by treatment of 2,2-dibromoacetophenone with morpholine followed by acidic hydrolysis.[10] Excellent yields of phenylglyoxal hemihydrate can be obtained on a small scale by the hydrolysis of phenylglyoxal hemimercaptal with boiling dilute hydrochloric acid [11] or in one step from 2-(methylsulfinyl)acetophenone by hydrolysis with boiling 8% phosphoric acid.

4. Merits of the Preparation

This procedure provides a convenient synthesis of phenylglyoxal from readily available starting materials. In addition, the method described appears to have general utility for the synthesis of glyoxals. It has been used for the synthesis of *p*-tolylglyoxal, *p*-methoxyphenylglyoxal, *p*-bromophenylglyoxal, and cyclohexylglyoxal. Since β-keto sulfoxides are readily alkylated in basic solution to yield α-alkyl β-keto sulfoxides,[12] it would appear possible to extend the scope of the reaction to yield a variety of α-diketones.

1. Department of Chemistry, Iowa State University, Ames, Iowa 50010.
2. H. von Pechmann, *Ber.*, **20**, 2904 (1887); H. Müller and H. von Pechmann, *Ber.*, **22**, 2556 (1889); A. Pinner, *Ber.*, **35**, 4131 (1902); **38**, 1531 (1905); I. Smedley, *J. Chem. Soc.*, **95**, 218 (1909).
3. C. Neuberg and E. Hofmann, *Biochem. Z.*, **229**, 443 (1930).
4. C. Neuberg and E. Hofmann, *Biochem. Z.*, **239**, 495 (1931); S. Cusmano, *Gazz. Chim. Ital.*, **68**, 129 (1938).
5. J. U. Nef, *Ann.*, **335**, 269 (1904); M. Henze, *Z. Physiol. Chem.*, **198**, 82 (1931); **200**, 232 (1931).
6. W. Madelung and M. E. Oberwegner, *Ber.*, **65**, 931 (1932).
7. H. L. Riley, J. F. Morley, and N. A. C. Friend, *J. Chem. Soc.*, 1875 (1932); H. L. Riley, Brit. Patent 354,798 (1930) [*C.A.*, **26**, 3804 (1932)]; U.S. Patent 1,955,890 (1934) [*C.A.*, **28**, 4067 (1934)]; H. A. Riley and A. R. Gray, *Org.*

Syntheses, Coll. Vol., **2**, 509 (1943); R. Bousset, *Bull. Soc. Chim. France*, [5] **6**, 986 (1939).
8. N. Kornblum, J. W. Powers, G. J. Anderson, W. J. Jones, H. O. Larson, O. Levand, and W. M. Weaver, *J. Am. Chem. Soc.*, **79**, 6562 (1957).
9. F. Weygand and H. J. Bestmann, *Ber.*, **90**, 1230 (1957).
10. M. Kerfanto, *Compt. Rend.*, **254**, 493 (1962).
11. H.-D. Becker, G. J. Mikol, and G. A. Russell, *J. Am. Chem. Soc.*, **85**, 3410 (1963).
12. E. J. Corey and M. Chaykovsky, *J. Am. Chem. Soc.*, **86**, 1639 (1964); **87**, 1345 (1965).

2-PHENYLINDAZOLE

(2*H*-Indazole, 2-phenyl-)

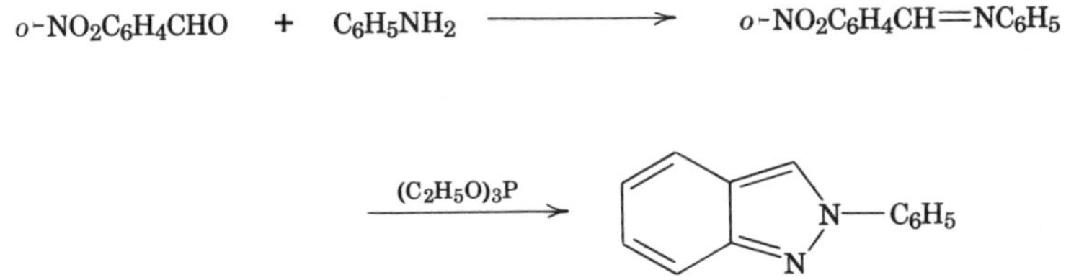

Submitted by J. I. G. Cadogan and R. K. Mackie [1]
Checked by William G. Dauben, Harold B. Morris, and Kent E. Opheim

1. Procedure

A. *o-Nitrobenzalaniline.* A mixture of 14 g. (0.15 mole) of aniline (Note 1) and 22.7 g. (0.15 mole) of *o*-nitrobenzaldehyde (Note 2) is heated in a 100-ml. round-bottomed flask on a water bath for 1 hour, allowed to cool, and dissolved in 100 ml. of ether. The ethereal solution is dried, and the ether is removed by distillation. The residue solidifies on standing (Note 3) and is recrystallized from 55 ml. of water-ethanol (1:8) to yield 29.4–31.8 g. (87–94%) of yellow *o*-nitrobenzalaniline, m.p. 64–66° (Note 4).

B. *2-Phenylindazole.* In a 200-ml. round-bottomed flask fitted with a condenser are mixed 50 g. (0.30 mole) of triethyl phosphite (Note 5) and 22.6 g. (0.10 mole) of *o*-nitrobenzalaniline. The apparatus is sealed from the atmosphere by means of a liquid paraffin bubbler that consists of a U-tube the bend of which is just filled with mineral oil. The apparatus is flushed with nitrogen, and the contents are kept under nitrogen during the reaction. The mixture is heated at 150° in an oil bath for 8 hours and cooled, and the condenser is replaced by a Claisen distillation head. Triethyl phosphite, b.p. 46–48° (10 mm.), and triethyl phosphate, b.p. 90–92° (10 mm.), are removed by distillation under reduced pressure; the volume of distillate is 48–50 ml. On cooling, the black residue solidifies. The flask is filled with glass wool, and the remaining phosphite and phosphate (1–3 g.) are removed by distillation at 30–50° (1 mm.). The residue of crude 2-phenylindazole is distilled at 10^{-4} mm.; b.p. 108–112°. The yield is 13–15 g. (67–78%) (Notes 6 and 7).

This product is crystallized from 75–100 ml. of ethanol-water (7:3) to yield pale yellow crystals, m.p. 81–82°. Additional material is obtained by dilution of the mother liquor with *ca.* 200 ml. of water and two crystallizations as before. The total yield is 10–12 g. (52–62%).

2. Notes

1. Aniline is purified by distillation from zinc dust.
2. The reagent as supplied by British Drug Houses or Eastman Organic Chemicals was used directly.
3. If the product does not solidify at room temperature, it should be cooled with dry ice.
4. *o*-Nitrobenzalaniline is very photosensitive and should be kept away from light as much as possible.
5. The reagent as supplied by Albright and Wilson, Ltd., or Matheson, Coleman and Bell was fractionally distilled from sodium and used within a few days of distillation.
6. A slightly purer sample may be obtained by chromatography on alumina. Elution with chloroform-benzene (1:4) gives a pale yellow solid which is purified further by crystallization from 70% ethanol.
7. The checkers found it more convenient to transfer the crude, black 2-phenylindazole to an apparatus for simple bulb-to-bulb distillation and not to retain the distillation head.

3. Methods of Preparation

The procedure given here is essentially that described previously by the submitters [2] and is based on the early work of Knoevenagel.[3] 2-Phenylindazole has been prepared by reduction of N-(*o*-nitrobenzyl)aniline with tin and hydrochloric acid,[4] by reduction of N-(*o*-nitrobenzyl)-N-nitrosoaniline with tin and hydrochloric acid,[5] by dehydration of 2-(phenylazo)benzyl alcohol,[6] by elimination of acetic acid from 2-(phenylazo)benzyl acetate,[7] by dehydrogenation of 3,3a,4,5,6,7-hexahydro-2-phenylindazole with sulfur,[8] and by thermal decomposition of *o*-azidobenzalaniline.[9]

4. Merits of the Preparation

Reductive cyclization of nitro compounds by triethyl phosphite is a general method for the preparation of a variety of nitrogen-containing heterocyclic systems. The submitters have synthesized the following ring systems by this method from the starting materials given in parentheses: 2-arylindoles (*o*-nitrostilbenes),[2] 2-arylindazoles (*o*-nitrobenzalanilines),[2] 2-arylbenzotriazoles (*o*-nitroazobenzenes),[2] carbazoles (*o*-nitrobiphenyls),[2] phenothiazines (*o*-nitrodiphenyl sulfides),[10, 11] and anthranils (*o*-nitrophenyl ketones).[10]

The products are isolated in good yield in a one-stage synthesis from starting materials that are readily available in the main. An alternative method involves the decomposition of the corresponding azides.[9, 12] These compounds are less readily available and are more hazardous to use than are the nitro compounds used in the present synthesis. This synthesis also gives better yields than the cyclization using ferrous oxalate,[12, 13] which is performed under much harsher conditions.

1. Department of Chemistry, St. Salvator's College, University of St. Andrews, St. Andrews, Fife, Scotland.
2. J. I. G. Cadogan, M. Cameron-Wood, R. K. Mackie, and R. J. G. Searle, *J. Chem. Soc.*, 4831 (1965); J. I. G. Cadogan and R. J. G. Searle, *Chem. Ind. (London)*, 1282 (1963).
3. E. Knoevenagel, *Ber.*, **31**, 2609 (1898).
4. C. Paal and F. Krecke, *Ber.*, **23**, 2634 (1890); C. Paal, *Ber.*, **24**, 959 (1891).
5. M. Busch, *Ber.*, **27**, 2897 (1894).

6. P. Freundler, *Compt. Rend.*, **136**, 1136 (1903); *Bull. Soc. Chim. France*, [3] **29**, 742 (1903).
7. P. Freundler, *Compt. Rend.*, **138**, 1425 (1904); *Bull. Soc. Chim. France*, [3] **31**, 868 (1904).
8. I. I. Grandberg, A. N. Kost, and L. S. Yaguzhinskii, *Zh. Obshch. Khim.*, **29**, 2537 (1959); *J. Gen. Chem. USSR* (*Engl. Transl.*), **29**, 2499 (1959).
9. L. Krbechek and H. Takimoto, *J. Org. Chem.*, **29**, 1150 (1964).
10. J. I. G. Cadogan, R. K. Mackie, and M. J. Todd, *Chem. Commun.*, 491 (1966).
11. J. I. G. Cadogan, S. Kulik, and M. J. Todd, unpublished results.
12. R. A. Abramovitch and B. A. Davis, *Chem. Rev.*, **64**, 149 (1964).
13. R. A. Abramovitch, Y. Ahmad, and D. Newman, *Tetrahedron Letters*, 752 (1961).

2-PHENYLPERFLUOROPROPENE

[Styrene, β,β-difluoro-α-(trifluoromethyl)-]

$$C_6H_5COCF_3 + CClF_2CO_2Na + (C_6H_5)_3P \rightarrow C_6H_5C(CF_3){=}CF_2 + (C_6H_5)_3PO + NaCl + CO_2$$

Submitted by Frank E. Herkes and Donald J. Burton [1]
Checked by W. C. Ripka and R. E. Benson

1. Procedure

Caution! This compound is an analog of the toxic olefin, perfluoroisobutylene. Since its toxicity is unknown, proper care should be exercised in its handling.

A 1-l., three-necked, round-bottomed flask is equipped with a mechanical stirrer, a nitrogen-inlet tube, and a reflux condenser connected to a dry ice-trichloroethylene trap (Note 1) that is followed by a water trap to measure carbon dioxide evolution (Note 2). The flask is charged with 65.6 g. (0.250 mole) of triphenylphosphine (Note 3) and 43.5 g. (0.250 mole) of α,α,α-trifluoroacetophenone (Note 4) in 200 ml. of dry diglyme (Note 5). The solution is heated to a bath temperature of 140° under a nitrogen atmosphere, and the nitrogen-inlet tube is replaced by a pressure-equalizing dropping funnel (Note 6) containing a solution of 76.2 g. (0.500 mole) of sodium chlorodifluoroacetate (Note 7) in 150 ml. of dry diglyme. The solution of sodium chlorodifluoroacetate is added dropwise over a period of 1 hour (Note 8), and the reaction mixture is heated for an additional hour at 130–140° to ensure complete decarboxylation of the salt (Note 9). The colors of the reaction mixture are characteristic of ylid reactions, changing from a creamy white to a creamy orange and finally to a deep brown.

The reaction mixture is then cooled to room temperature under nitrogen. The cool mixture is steam distilled until 2 l. of distillate has been collected. The lower, organic layer is separated from the distillate, washed with cold water (Note 10) to remove diglyme, and dried over anhydrous calcium sulfate. Fractional distillation gives 26–31 g. (50–60%) of 2-phenylperfluoropropene, b.p. 58–59° (54 mm.), $n^{20}{}_{D}$ 1.4225 (Notes 11 and 12).

2. Notes

1. In addition to carbon dioxide, small amounts of fluoroformyl fluoride and chloride are formed and swept out with the carbon dioxide.

2. The rate of decarboxylation can be followed qualitatively by collecting the liberated carbon dioxide over water, *e.g.*, by the use of a wet test meter.

3. Eastman Organic Chemicals white label triphenylphosphine was used directly.

4. Available from Pierce Chemical Co., Rockford, Illinois. The ketone can also be prepared conveniently from phenylmagnesium bromide and trifluoroacetic acid by the method of Dishart and Levine.[2]

5. Diglyme (Ansul Ether 141) was predried over calcium hydride and distilled under reduced pressure from lithium aluminum hydride; b.p. 62–63° (15 mm.).[3] The yield of olefin is dependent on the dryness of the solvent. The formation of 2-phenyl-2H-perfluoropropane is favored by the presence of water in the solvent.

6. A dropping funnel of the type described by Benson and McKusick [4] is satisfactory.

7. Sodium chlorodifluoroacetate is prepared in quantitative yield by careful neutralization of 130.5 g. (1.00 mole) of chlorodifluoroacetic acid (available from Allied Chemical Corp.) in 300 ml. of ether with 53.0 g. (0.500 mole) of anhydrous sodium carbonate, removal of the solvent and water under reduced pressure, and drying over phosphorus pentoxide in a vacuum desiccator. Studies have shown that the best yield of olefin is obtained when a 100% excess of salt is used.

8. The salt is added at a rate sufficient to cause a constant evolution of carbon dioxide. The reaction is slightly exothermic (*ca.* 10° temperature rise).

9. A total of 9460 ml. (72% STP) of carbon dioxide was collected.

10. Eight washings with 50-ml. portions of water were found to remove all the diglyme.

11. The submitters used an 18-in. spinning-band column. The product was shown to be >99.9% pure by gas-liquid chromatography on Carbowax 20M.

12. The checkers used a 40-cm. spinning-band column. The product, n^{25}_D 1.4237, was shown to be 98.9% pure by gas-liquid chromatography on a 6-ft. 20% fluorosilicone column. The retention time was 3.75 minutes with a flow rate of helium of 100 ml. per minute, and a column temperature of 125° with the injection port at 170°. The ^{19}F n.m.r. spectrum (56.4 MHz) consists of four lines of equal intensity centered at +3396 Hz from trichlorofluoromethane (internal) and two sets of two overlapping quartets centered at +4369 Hz and +4461 Hz, respectively. The integrated intensities of the three sets of fluorine resonances are 3:1:1.

3. Methods of Preparation

This procedure is a modification of the method previously reported by the submitters.[5] 2-Phenylperfluoropropene has been reported as a by-product of the thermal decomposition of 7,7-bis-(trifluoromethyl)-1,3,5-cycloheptatriene; however, no experimental procedure was given.[6]

4. Merits of the Preparation

The procedure illustrates a fairly general method for the preparation of β-substituted perfluoroolefins. The method has been applied to the synthesis of 2-cyclohexyl- (70%), 2-benzyl- (61%), and 2-(*p*-fluorophenyl)perfluoropropenes (67%), and it is probably applicable to any α-trifluoromethyl ketone. Olefins containing a perfluoroalkyl group other than trifluoromethyl can be prepared by the same procedure by the substitution of lithium chlorodifluoroacetate for sodium chlorodifluoroacetate.[7] Other routes to β-substituted perfluoroolefins are not general or convenient. Routes to perfluoroolefins generally yield the α-substituted olefin rather than the β-substituted olefin.

This method can also be utilized as a general method for the preparation of olefins with terminal difluoromethylene groups from aldehydes.[8] Also, by the substitution of tributylphosphine for triphenylphosphine in this procedure, ketones other than those containing an α-perfluoroalkyl group can be converted to terminal difluoromethylene compounds.[9]

1. Department of Chemistry, University of Iowa, Iowa City, Iowa 52240.

2. K. T. Dishart and R. Levine, *J. Am. Chem. Soc.*, **78**, 2268 (1956).
3. G. Zweifel and H. C. Brown, *Org. Reactions*, **13**, 28 (1963).
4. R. E. Benson and B. C. McKusick, *Org. Syntheses*, Coll. Vol. **4**, 747 (1963).
5. D. J. Burton and F. E. Herkes, *Tetrahedron Letters*, 1883 (1965).
6. D. M. Gale, W. J. Middleton, and C. G. Krespan, *J. Am. Chem. Soc.*, **87**, 657 (1965).
7. D. J. Burton and F. E. Herkes, *Tetrahedron Letters*, 4509 (1965).
8. S. A. Fuqua, W. G. Duncan, and R. M. Silverstein, *J. Org. Chem.*, **30**, 1027 (1965); *Org. Syntheses*, **47**, 49 (1967).
9. S. A. Fuqua, W. G. Duncan, and R. M. Silverstein, *J. Org. Chem.*, **30**, 2543 (1965).

(+)- AND (−)-α-(2,4,5,7-TETRANITRO-9-FLUORENYLIDENEAMINOOXY)PROPIONIC ACID

(Propionic acid, 2-(2,4,5,7-tetranitrofluoren-9-ylideneaminooxy)-, (+)- and (−)-)

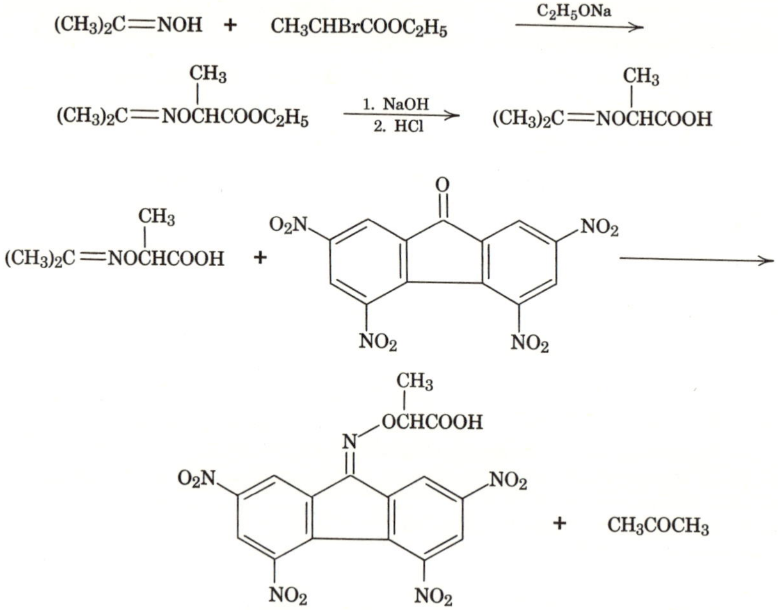

Submitted by Paul Block, Jr.,[1] and Melvin S. Newman [2]
Checked by William G. Dauben, Milton E. Lorber,
Carroll S. Montgomery, and Gary W. Shaffer

1. Procedure

A. *Ethyl α-(isopropylideneaminooxy)propionate.* A 1-l. three-necked flask is equipped with a mechanical stirrer, a dropping funnel, and a thermometer that can be immersed in the contents of the flask. There is added to the flask 500 ml. of commercial absolute ethanol (Note 1) followed by 17.5 g. (0.76 g. atom) of sodium, which is added carefully in small portions. When a clear solution has been obtained, 55.0 g. (0.75 mole) of acetone oxime is added (Note 2). The flask is cooled in a water bath held at 5–10°, the stirrer is started, and 136 g. (0.75 mole) of ethyl α-bromopropionate (Note 3) is added during 20–30 minutes at a rate such that the temperature does not rise above 20°. The cooling bath is removed, and the stirring is continued until the contents of the flask reach room temperature.

After standing for 12 hours, the reaction mixture is filtered by gravity into a 1-l. round-bottomed flask, and the solid sodium bromide is washed with 50 ml. of ethanol. The combined filtrate and washings are concentrated to a volume of about 400 ml., and 250 ml. of water is added to the cooled concentrate. The mixture is extracted with 50 ml. of a 1:1 mixture of ether and benzene, and the aqueous layer is reextracted with 100 ml. of the same solvent mixture. The organic extracts are combined, washed with 100 ml. of water and 50 ml. of saturated aqueous sodium chloride, and filtered through a few grams of anhydrous magnesium sulfate. The solvent is removed on a rotary evaporator, and the residue is distilled to yield 71–77 g. (55–59%) of ethyl α-(isopropylideneaminooxy)propionate, b.p. 62–64° (4 mm.).

B. *d,l-α-(Isopropylideneaminooxy)propionic acid.* In a 1-l. three-necked flask fitted with a stirrer and a thermometer that can be immersed in the contents of the flask is placed 300 ml. of 5% aqueous sodium hydroxide (0.37 mole). The flask is heated on a water bath until the temperature of the solution reaches 70°, and 52 g. (0.30 mole) of ethyl α-(isopropylideneaminooxy)propionate is added. The mixture is stirred rapidly while the temperature is held at 70°; the stirring is continued for 20 minutes beyond the time necessary for the contents of the flask to become homogeneous (Note 4). The solution is cooled and acidified to Congo red paper with 5*N* hydrochloric acid, and 175 g. of ammonium sulfate is added. The mixture is extracted three times with a total of 300 ml. of a 1:1 mixture of ether and benzene. The combined extracts are dried rapidly over 5 g. of anhydrous magnesium sulfate and filtered (Note 5). The solvent is removed by distillation, and 160 ml. of petroleum ether (b.p. 30–60°) is added to the cooled residue. The resulting solution is placed in a refrigerator for several hours. The crystals that separate are removed by suction filtration and washed with a small volume of cold petroleum ether. The yield of colorless product is 35–37 g. (80–85%); m.p. 59–60.5° (Note 6).

C. (+)- and (−)-*α-(Isopropylideneaminooxy)propionic acid-(−)-ephedrine salts.* A solution of 36.6 g. (0.200 mole) of (−)-ephedrine monohydrate (Note 7) in 800 ml. of ethyl acetate containing 6% of ethanol (Note 8) is placed in a 2-l. beaker. *d,l-α-*(Isopropylideneaminooxy)propionic acid (29.0 g., 0.200 mole) is dissolved in this solution by stirring (Note 9). The beaker is covered securely with a rubber dam, cooled for a short period in an ice bath, placed in a refrigerator at 0–5°, and allowed to remain undisturbed for 8–16 hours after crystallization has begun (Note 10). The solid mass of crystals is filtered by suction, and the funnel is covered with a rubber dam to remove most of the solvent. The solid product is placed in a 500-ml. beaker, 250 ml. of ethyl acetate is added (Note 11), and the mixture is heated until all the solid has dissolved. The solution is cooled, placed in a refrigerator for several hours, and filtered; the crystalline precipitate is dried in air. The yield of the (−)-ephedrine-(+)-α-(isopropylideneaminooxy)propionic acid salt is 22–25 g. (71–81%); m.p. 115–119° (Notes 12 and 13); $[\alpha]^{20}\text{D}$ −4.2° (*c* 1.5, chloroform).

The combined filtrates are diluted with an equal volume of petroleum ether (b.p. 30–60°), placed in a refrigerator for 8–16 hours, and filtered. The solid product is recrystallized from ethyl acetate (10 ml. per gram of the salt). The yield of the monohydrate of the (−)-ephedrine-(−)-α-(isopropylideneaminooxy)propionic acid salt is 19–26 g. (58–79%); m.p. 88–90°; $[\alpha]^{20}\text{D}$ −57° (*c* 1.5, chloroform) (Notes 13 and 14).

D. (+)- *and* (−)-*α-(Isopropylideneaminooxy)propionic acid.* To a solution of 20 g. (0.064 mole) of the (−)-base-(+)-acid salt in 60 ml. of water is added 14 ml. (0.070 mole) of 5*N* hydrochloric acid. The solution is filtered to remove a slight insoluble residue and extracted with four 25-ml. portions of a 1:1 mixture of ether and benzene. The combined extracts are dried rapidly over

1–2 g. of anhydrous magnesium sulfate and filtered. The organic solvents are removed by distillation from a steam bath, the residue is dissolved in 75 ml. of petroleum ether (b.p. 30–60°), and the solution is allowed to stand in a refrigerator for 12 hours. The crystalline product (7.0–7.5 g.; m.p. 75–81°) is collected and dissolved in hot acetone (0.5 ml. per gram), and the solution is diluted with hexane (5 ml. per gram). The solution is placed in a refrigerator for 8–16 hours, and the crystalline (+)-α-(isopropylideneaminooxy)propionic acid (5.5–6.5 g.; 59–70%) that separates is collected; m.p. 83–85°; $[\alpha]^{20}\text{D}$ +32° (*c.* 1.6, water).

In a similar manner, from 20 g. (0.061 mole) of the monohydrate of the (−)-base-(−)-acid salt, there is obtained 6.4–6.7 g. (73–76%) of the (−)-acid, m.p. 83–85°, $[\alpha]^{20}\text{D}$ −29° (*c* 1.44, water), directly from the crystallization from petroleum ether. Subsequent recrystallization from acetone-hexane is normally not required.

E. (+)- *and* (−)-*α-(2,4,5,7-Tetranitro-9-fluorenylideneaminooxy)propionic acid* (*TAPA*). To a solution of 5.5 g. (0.038 mole) of either optical antipode of α-(isopropylideneaminooxy)-propionic acid in 85 ml. of 96% acetic acid in a 250-ml. round-bottomed flask are added 9.0 g. (0.025 mole) of 2,4,5,7-tetranitrofluorenone,[3] 0.30–0.35 ml. of concentrated sulfuric acid, and a few boiling chips. The flask is fitted with an air condenser (Note 15), and the contents are heated under reflux so that the condensing liquid nearly reaches the top of the condenser (Note 16). After 2 hours, 18 ml. of water is added to the hot solution, and crystallization is allowed to take place slowly, first at room temperature and finally for 12 hours in a refrigerator. The yellow crystalline acid is filtered and dissolved in 70 ml. of hot acetic acid. The solution is diluted while hot with 60 ml. of water, cooled rapidly with stirring, and kept at 0° for several hours. The optically active TAPA is filtered and air-dried away from direct sunlight until the odor of acetic acid is negligible. The crystals are then dried in an oven at 110° (Note 17) and protected from light by storage in a suitable container; yield 7.8–10.0 g. (70–90%). The TAPA from the (−)-acid has $[\alpha]^{25}\text{D}$ +97° and that from the (+)-acid $[\alpha]^{25}\text{D}$ −97° (Note 18).

2. Notes

1. Pure anhydrous ethanol [4] offers no advantage over commercial absolute ethanol.

2. "Eastman grade" acetone oxime was used as obtained from Eastman Organic Chemicals.

3. "Eastman grade" ethyl α-bromopropionate was used as obtained from Eastman Organic Chemicals.

4. Usually 10–20 minutes are required to obtain complete reaction.

5. If the solution is not entirely colorless, it should be shaken with a small amount of activated carbon and filtered before distillation.

6. The checkers found 53–56°; m.p. 57–61° has been reported.[5]

7. "Ephedrine alkaloid hydrous," Merck, was used. If anhydrous ephedrine is employed, only 33 g. should be used, and 3.6 g. (0.20 mole) of water should be added. Anhydrous conditions lead to incomplete resolution.

8. Commercial absolute ethanol (48 ml.) is pipetted into a 1-l. graduated cylinder and diluted with 800 ml. of ethyl acetate ("Eastman grade").

9. Both components are soluble in ethyl acetate at room temperature; the resulting salt is not. By dissolving the components sequentially, precipitation of the salt is generally avoided. Should the salt form, however, it must be dissolved by gentle heating.

10. Prolonged standing must be avoided as the deposition of the (−)-ephedrine-(−)-acid salt can occur.

11. Ethanol is not added to the ethyl acetate at this point.

12. Highly purified samples have m.p. 124.0–124.5°.

13. The two diastereoisomeric salts can be readily distinguished from each other. The (−)-ephedrine-(+)-acid salt is formed as cottony crystals that grow in the solution and eventually become a solid, white opaque mass. The monohydrate of the (−)-ephedrine-(−)-acid salt consists of clear, chunky crystals that grow from, and adhere to, the bottom and sides of the flask.

14. The water of hydration is lost on standing in a desiccator

over phosphorus pentoxide; the melting point eventually reached is 109–110°.[6]

15. A 250 × 15-mm. glass tube is satisfactory.

16. The suspended tetranitrofluorenone dissolves completely in about 25 minutes; the vigorous heating is required to bring about the solution and reaction.

17. One mole of acetic acid of solvation is lost only slowly at room temperature; the solvated product has m.p. *ca.* 123°.[5] The submitters found that the air-dried material, on being dried at 110°, yielded essentially solvent-free compound, m.p. 201–203° (dec. with prior darkening). The checkers found that at 110° the air-dried material melted, turned brown, and then resolidified. They also found that the material, on being dried at 70–80° (1 mm.) over potassium hydroxide pellets for several days, remained yellow but melted over a range 110–125°, resolidified, and remelted at 190–195°.

18. The checkers used material dried at 70–80° for their determination of the rotation and obtained values in agreement with those reported by the submitters.

3. Methods of Preparation

TAPA has been prepared only as described in this procedure.[5] α-(Isopropylideneaminooxy)propionic acid has been prepared and resolved by the present procedure [6] and has been prepared directly from α-bromopropionic acid and resolved as the (−)-ephedrine salt by crystallization from hydrocarbon mixtures.[5]

4. Merits of the Preparation

The use of ethyl α-bromopropionate simplifies the preparation of α-(isopropylideneaminooxy)propionic acid. Resolution in ethyl acetate solution has proved less erratic than in the hydrocarbon solvents previously recommended,[5] and the isolation of both diastereoisomeric salts formed is facilitated. TAPA has found use in the resolution of polycyclic aromatic compounds that do not possess functional groups that would permit resolution by other methods.[5, 7, 8]

1. Department of Chemistry, University of Toledo, Toledo, Ohio.
2. Department of Chemistry, The Ohio State University, Columbus, Ohio 43210.
3. M. S. Newman and H. Boden, *Org. Syntheses*, **42**, 95 (1962).
4. R. H. Manske, *Org. Syntheses*, Coll. Vol. **2**, 154 (1943).
5. M. S. Newman and W. B. Lutz, *J. Am. Chem. Soc.*, **78**, 2469 (1956).
6. P. Block, Jr., *J. Org. Chem.*, **30**, 1307 (1965).
7. M. S. Newman, R. G. Mentzer, and G. Slomp, *J. Am. Chem. Soc.*, **85**, 4018 (1963).
8. M. S. Newman, R. W. Wotring, Jr., A. Pandit, and P. M. Chakrabarti, *J. Org. Chem.*, **31**, 4293 (1966).

2,2,4-TRIMETHYL-3-OXOVALERYL CHLORIDE

(Valeryl chloride, 2,2,4-trimethyl-3-oxo-)

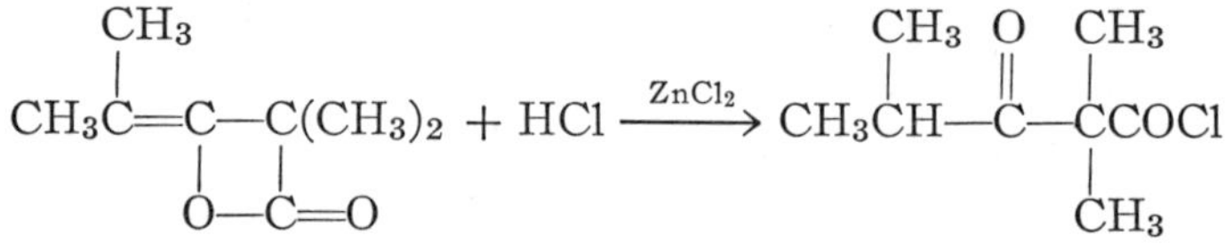

Submitted by EDWARD U. ELAM, P. GLENN GOTT, and ROBERT H. HASEK[1]
Checked by V. BOEKELHEIDE and G. SINGER

1. Procedure

Caution! The starting material in this preparation, 3-hydroxy-2,2,4-trimethyl-3-pentenoic acid β-lactone, is a mild but deceptively persistent lachrymator.

A mixture of 5 g. of anhydrous zinc chloride (Note 1) and 280 g. (2.00 moles) of 3-hydroxy-2,2,4-trimethyl-3-pentenoic acid β-lactone[2] is placed in a 500-ml. three-necked flask equipped with a sealed stirrer (Note 2), a coarse fritted-glass gas-dispersion thimble, a thermometer immersed in the liquid, and an air-cooled reflux condenser (Note 3). The outlet of the condenser is connected to a bubble counter filled with concentrated sulfuric acid; this in turn is vented to the atmosphere through a water scrubber. The flask is immersed in an ice bath, and stirring is started. When the temperature of the mixture is about 10°, anhydrous hydrogen chloride is introduced through the gas-dispersion tube at such a rate that a low stream of bubbles escapes through the bubble counter. Gas absorption is slow at first, but after a few minutes the zinc chloride dissolves and the rate of gas absorption increases sharply. The temperature rises rapidly to 50–70° at the same time (Note 4). After about 10 minutes the temperature falls and the rate of gas absorption decreases. The ice bath is removed, and the addition of hydrogen chloride is continued for 30 minutes.

The reaction mixture is distilled rapidly under reduced pressure through a short Vigreux column (Note 5). The yield of crude product, b.p. 50–80° (5–10 mm.), n^{20}_{D} 1.4410–1.4416, is 310–330 g. (88–93%). This material is sufficiently pure for most purposes (Note 6). Fractionation through a 1 × 36-in. column packed with 8 × 8-mm. glass helices gives, after removal of a small amount of forerun, pure 2,2,4-trimethyl-3-oxovaleryl chloride, b.p. 86° (23 mm.), n^{20}_{D} 1.4418.

2. Notes

1. Reagent grade anhydrous zinc chloride from a freshly opened bottle may be used without special drying.

2. A glass or Teflon stirrer should be used.

3. An air-cooled condenser long enough to trap escaping spray is sufficient since the mixture is never hot enough to reflux.

4. Overheating of the mixture at this point or during the subsequent distillation causes decomposition of the crude acid chloride with formation of tarry by-products. This decomposition in the presence of zinc chloride is fairly rapid at temperatures above 100°.

5. Since hydrogen chloride is evolved in the early stages of this distillation, no effort is made to control the pressure. The vacuum pump is protected from the hydrogen chloride by insertion of a 1.5 × 15-in. glass tube packed with sodium hydroxide pellets in the vacuum line. The purpose of this distillation is to remove dissolved hydrogen chloride and the zinc chloride catalyst; it should be completed as rapidly as possible, under the best vacuum attainable, with no attempt at fractionation. The distillate may be fractionally distilled for further purification if

desired. 2,2,4-Trimethyl-3-oxovaleryl chloride is stable at its normal boiling point, 190° (730 mm.), after the zinc chloride has been removed. Contact of the acid chloride with metals should be avoided.

6. The chlorine content of the crude product varied from 19.9% to 20.1% in successive experiments; saponification analysis indicated a purity above 97%.

3. Methods of Preparation

The procedure described is related to that for preparing acetoacetyl chloride from diketene and hydrogen chloride.[3]

4. Merits of the Preparation

Unlike acetoacetyl chloride, which decomposes at temperatures above −20°, 2,2,4-trimethyl-3-oxovaleryl chloride is stable at elevated temperatures. It may find use as an intermediate; for example, it can be used in the preparation of acid chlorides by an exchange reaction which is forced to completion by decarboxylation of the by-product, β-keto acid:

$$RCO_2H + (CH_3)_2CHCOC(CH_3)_2COCl \rightleftharpoons RCOCl + (CH_3)_2CHCOC(CH_3)_2CO_2H$$

$$\downarrow$$

$$(CH_3)_2CHCOCH(CH_3)_2 + CO_2$$

1. Research Laboratories, Tennessee Eastman Co., Kingsport, Tennessee 37662.
2. R. H. Hasek, R. D. Clark, and G. L. Mayberry, *Org. Syntheses*, this volume, p. 72.
3. C. D. Hurd and C. D. Kelso, *J. Am. Chem. Soc.*, **62**, 1548 (1940).

VINYL TRIPHENYLPHOSPHONIUM BROMIDE

(Phosphonium bromide, triphenylvinyl-)

$$C_6H_5OCH_2CH_2Br + (C_6H_5)_3P \xrightarrow{C_6H_5OH}$$

$$C_6H_5OCH_2CH_2\overset{+}{P}(C_6H_5)_3\ Br^- \xrightarrow{CH_3CO_2C_2H_5}$$

$$CH_2{=}CH\overset{+}{P}(C_6H_5)_3\ Br^- + C_6H_5OH$$

Submitted by Edward E. Schweizer and Robert D. Bach[1]
Checked by François X. Garneau and Peter Yates

1. Procedure

Caution! Because phenol and its solutions are corrosive, rubber gloves should be used in the following operations. The product, vinyl triphenylphosphonium bromide, has been found to induce a sneezing, allergic reaction and contact with it should be avoided.

In a 1-l. three-necked flask is placed 1 lb. of reagent grade phenol (Note 1), 100 g. (0.50 mole) of β-bromophenetole, and 131 g. (0.50 mole) of triphenylphosphine (Note 2). The flask is equipped with a sealed stirrer, a thermometer, and a reflux condenser fitted with a calcium chloride drying tube. The mixture is stirred and heated to 90° ± 3° (Note 3) and kept at this temperature for 48 hours.

The solution is cooled to room temperature and added slowly (during 45 minutes) from a dropping funnel to vigorously stirred anhydrous ether (3 l.) in a 4-l. beaker. Material that adheres to the sides of the beaker is scraped down with a long steel spatula, and the mixture is filtered by suction. The solid product is transferred into a second 3-l. portion of anhydrous ether, and the mixture is stirred vigorously for 15 minutes and filtered by suction. The product is washed with three 250-ml. portions of warm anhydrous ether.

The white crystalline residue of crude phenoxyethyltriphenylphosphonium bromide (Note 4) is placed in a 3-l. two-necked flask equipped with a sealed stirrer and a reflux condenser fitted with a calcium chloride drying tube. Reagent grade (Note 5) ethyl acetate (1.5 l.) is added, and the solution is stirred under reflux for 24 hours. The mixture is cooled to room temperature, and the ethyl acetate layer is decanted (or filtered if the salt is crystalline). This procedure is repeated until the filtered salt, vinyl triphenylphosphonium bromide, melts at 186° and higher (Note 6).

After the final filtration the product is washed with two 100-ml. portions of ethyl acetate and two 100-ml. portions of anhydrous ether and dried for 24 hours at 80°. The dried, analytically pure vinyl triphenylphosphonium bromide, m.p. 186–190°, weighs 122–158 g. (66–86%).

2. Notes

1. The checkers found that it was important to use phenol free of colored impurities; in a run in which phenol, m.p. 39.5–41°, with a slight rose tinge was used the yield of product was reduced to 56%.

2. The β-bromophenetole was obtained from Aldrich Chemical Co.; the triphenylphosphine was obtained from M and T Chemicals, Inc., or Carlisle Chemical Works, Reading, Ohio, and recrystallized once from anhydrous ether (with filtration).

3. It is important that the temperature does not rise above 95°. There is a slight exotherm on initial heating that may necessitate the removal of the heating mantle in order to maintain a temperature below 95°.

4. Pure samples of this material may be obtained by using acetic acid as solvent instead of phenol.[2]

5. Because the vinyl salt reacts with ethanol, and decomposition of the phenoxyethyl precursor is inhibited by acetic acid,[2] reagent grade ethyl acetate is recommended.

6. Four treatments have always been necessary. The residue has always crystallized on cooling after the third treatment.

3. Methods of Preparation

The present procedure is that described by the submitters.[2] Vinyl triphenylphosphonium bromide has also been prepared by dehydrobromination of 2-bromoethyltriphenylphosphonium bromide, but no preparative details or yields have been disclosed.[3]

4. Merits of the Preparation

This salt has been used as a general reagent for the preparation of some heterocyclic and carbocyclic systems.[4–9] A variety of salts of type $XCH_2CH_2\overset{+}{P}(C_6H_5)_3\ Br^-$ has been prepared from the vinyl salt by treatment with alcohols, thiophenol, and diethylamine.[2]

1. Department of Chemistry, University of Delaware, Newark, Delaware 19711.
2. E. E. Schweizer and R. D. Bach, *J. Org. Chem.*, **29**, 1746 (1964).
3. D. Seyferth, J. S. Fogel, and J. K. Heeren, *J. Am. Chem. Soc.*, **86**, 307 (1964).
4. E. E. Schweizer, *J. Am. Chem. Soc.*, **86**, 2744 (1964).
5. E. E. Schweizer and K. K. Light, *J. Am. Chem. Soc.*, **86**, 2963 (1964).
6. E. E. Schweizer, L. D. Smucker, and R. J. Votral, *J. Org. Chem.*, **31**, 467 (1966).
7. E. E. Schweizer and G. J. O'Neill, *J. Org. Chem.*, **30**, 2082 (1965).
8. E. E. Schweizer and K. K. Light, *J. Org. Chem.*, **31**, 870 (1966).
9. E. E. Schweizer and L. D. Smucker, *J. Org. Chem.*, **31**, 3146 (1966).

SUBJECT INDEX

(This index comprises material from Volumes 45, 46, 47, and 48 only; for previous volumes see Collective Volumes 1 through 4 and Volume 44.)

Names in small capital letters refer to the titles of individual preparations. A number in ordinary boldface type denotes the volume. A page number in boldface italics indicates that the detailed preparative directions are given or referred to; entries so treated include principal products and major by-products, special reagents or intermediates (which may or may not be isolated), compounds mentioned in the text or Notes as having been prepared by the method given, and apparatus described in detail or illustrated by a figure. Page numbers in ordinary type indicate pages on which a compound or subject is mentioned in connection with other preparations.